Carlo Vezzoli · Fabrizio Ceschin
Lilac Osanjo · Mugendi K. M'Rithaa
Richie Moalosi · Venny Nakazibwe
Jan Carel Diehl

Designing Sustainable Energy for All

Sustainable Product-Service System Design Applied to Distributed Renewable Energy

With Mary Suzan Abbo, Elisa Bacchetti, Emanuela Delfino, Silvia Emili, Paulson Lethsolo, Mackay Okure, Yaone Rapitsenyane, Ephias Ruhode, James Wafula, Edurne Battista, Andrea Broom, Fiammetta Costa

Carlo Vezzoli
Design Department
Politecnico di Milano
Milan
Italy

Fabrizio Ceschin
College of Engineering, Design and Physical Sciences—Department of Design
Brunel University London
Uxbridge
UK

Lilac Osanjo
University of Nairobi
Nairobi
Kenya

Mugendi K. M'Rithaa
Cape Peninsula University of Technology
Cape Town
South Africa

Richie Moalosi
University of Botswana
Gaborone
Botswana

Venny Nakazibwe
The College of Engineering, Design, Art and Technology
Makerere University
Kampala
Uganda

Jan Carel Diehl
Delft University of Technology
Delft
The Netherlands

https://doi.org/10.1007/978-3-319-70223-0

Foreword I

It gives me great honour and privilege to contribute with the foreword for this book, as a collaboration between Africa and Europe towards sustainable energy for All. The book is a transdisciplinary text with the latest knowledge-base, know-how and experiences in sustainable energy for All system development and design. Indeed, it presents the key role of design in providing sustainable energy solutions to human society in its quest for continuous improvement and socioeconomic development.

Africa is blessed with a variety of sustainable energy resources, including solar, hydropower, wind, mini/micro hydro and geothermal resources. However, lack of access to adequate and sustainable energy services remains one of the major constraints to economic development on the continent. We, therefore, urgently need to encourage and incentivize our scientists, researchers and research institutions, businesses and industries, supported by development partners and governments, to invest more in research, education, curricula development and designing feasible and bankable renewable energy projects, that will enable Africa to exploit and utilise the continent's vast energy resources for social economic development. From the LeNSes project and from this book, a proposal was developed in this direction. Indeed, the Sustainable Product-Service System (S.PSS) model applied to the Distributed Renewable Energy (DRE) one, a promising win-win combined model towards sustainable energy for All, in fact, promoting the leapfrog from individual ownership of energy systems to collective access to sustainable energy. To design and implement such new models, the book presents the knowledge-base, the design approach, the design process and related tools developed and tested during the LeNSes project.

It is my opinion that this book can effectively contribute to promote a win-win approach towards sustainable energy solutions, which has been developed in the African continent, and can be replicated elsewhere for the universal attainment of sustainable energy for All.

Kampala, Uganda

Irene Muloni
Minister of Energy and Mineral Development

Foreword II

During the last few decades, the history of design culture and practice, when dealing with the issue of sustainability, has moved from individual products to systems of consumption and production, and from strictly environmental problems to the complex blend of socioethical, environmental and economic issues. Even more recently a new challenge becomes very clear: Sustainable Energy for All (accessible even to low- and middle-income people) is a key leverage for sustainable development, with both environmental and socioethical benefits.

Within this framework, it is key important that design can take a proactive role and become an agent to extend the access to sustainability energy. It can do so because within its genetic code there is the idea that its role is to improve the quality of the world: an ethical–cultural component that, though not generally apparent, can be found in a deeper examination of the majority of designers' motivations.

Finally, it is far obvious that a key role has to be played by the Higher Education Institutions, both in researching and defining the new roles the designers may play, as well as in the curricular proposal where a new generation of design should grow.

A challenging journey is ahead of us. And from this perspective we believe this book will contribute to a larger change in the design community requested to meet this challenge.

Milan, Italy

Silvia Piardi
Head of the Design Department
Politecnico di Milano

Foreword III

The sustainability framework has brought the use of renewable technologies to the fore. The publication of this book is another milestone in contributing towards the sustainability agenda. The book will serve as an interdisciplinary platform for sharing the latest knowledge and experiences in sustainable energy for practitioners, designers and researchers alike. It gives me a deep sense of gratitude that the University of Botswana through the Department of Industrial Design and Technology is contributing chapters, with state-of-the-art knowledge on the frontiers of system design for sustainable energy for all. The book has developed new methods of analysis and provides new solutions to keep up with the ever-changing frontiers of sustainable energy. I think that the authors can be confident that there will be many grateful readers who will gain a broader perspective of the disciplines of Design of Sustainable Product-Service System applied to Distributed Renewable Energies as a result of their efforts.

Gaborone, Botswana

Benjamin Bolaane
Professor and Dean, Faculty
of Engineering and Technology,
University of Botswana

Foreword IV

Access to renewable energy plays a crucial role in social and economic development, particularly in low- and middle-economy contexts. Technology is important but alone it is not the solution. The concept of Product-Service Systems (PSS) is a very promising approach to enable (renewable) energy technologies to contribute to a more sustainable society: PSS are able to integrate them within services and business models as well as to match renewable technologies with the needs and wishes of end users and other stakeholders. We as PSS developers and appliers are very pleased with the design-oriented and practical approach of this book, based on the rich experiences of all partners involved in the LeNSes project. We are convinced of and looking forward to the positive impact it will make, stimulating creative pathways to a more Sustainable Future.

Delft, The Netherlands/
Aalborg, Denmark

Prof. Dr. Ir. Han Brezet
Delft University of Technology;
Aalborg University

Foreword V

Ensuring energy access is one of the most important global challenges that we need to address. Currently, 1.2 million people lack access to electricity, with the majority of these living in rural areas in low- and middle-income countries. The impact of this energy poverty can be measured in acute respiratory illness caused by indoor pollution due to the use of kerosene and biomass (including dung) for lighting, heating and cooking. In addition, infant mortality is high due to the lack of refrigeration for vaccine and medicine storage and the inability to power incubators, and education is severely impacted.

The adoption of distributed renewable energy systems represents a promising strategy to tackle the problem. However, the challenge cannot be addressed by only considering the technological aspects. Innovative energy services and business models, and appropriate stakeholder value chains need also to be considered and coupled with appropriate technologies for energy generation, storage and transmission.

In this context, design, with its human-centred approach and technical competence, is well placed to play an important role as the technical and socio-cultural agent of change. This book explores how strategic design can provide a solution to the problem of universal energy access. It does so by proposing an innovative approach that focuses on the design of Product-Service Systems to deliver distributed renewable energy solutions that are economically viable and environmentally and socio-ethically sustainable. This book provides the theoretical foundations of this design approach, as well as an articulated set of methods and tools that can be used by practitioners and businesses.

This book is part of an overall effort that Brunel Design is undertaking to place research on global challenges at the core of the agenda. It also represents an excellent example of how research can quickly be translated into innovative design teaching programmes. As educators, we have a responsibility to ensure that our graduates are ethical, sustainable and responsible and understand their potential to make a meaningful contribution to societal well-being.

Uxbridge, UK

Dr. Ian de Vere
Head of Design Brunel University London

Preface I

The book is one of the outcomes of LeNSes, the Learning Network on Sustainable energy system, a project funded by the ACP-EU Co-operation Programme in Higher Education (EDULINK II), for curricula development and teaching diffusion in worldwide design higher education institutions, on design for sustainability focused on Sustainable Product-Service System (S.PSS) applied to Distributed Renewable Energies (DRE).

Milan, Italy	Carlo Vezzoli
Uxbridge, UK	Fabrizio Ceschin
Nairobi, Kenya	Lilac Osanjo
Cape Town, South Africa	Mugendi K. M'Rithaa
Gaborone, Botswana	Richie Moalosi
Kampala, Uganda	Venny Nakazibwe
Delft, The Netherlands	Jan Carel Diehl

Preface II

The twenty-first century has ushered in new technological capabilities to help ameliorate the plight of humanity. Notwithstanding, the wicked problems that designers, engineers and allied professionals grapple with have increased in complexity, scale and scope. In response to this emerging reality, the World Design Organization (WDO) has embraced a markedly transdisciplinary approach to inform the industrial design profession's efforts at promoting design for a better world. To advance this thinking, the WDO proffered a renewed definition stating that 'Industrial Design is a strategic problem-solving process that drives innovation, builds business success, and leads to a better quality of life through innovative products, systems, services, and experiences'. The WDO's renewed commitment to socially conscious design further strengthens the alignment with the empathic and inclusive philosophy of LeNSes as it relates to the Sustainable Product-Service System (S.PSS) model. Additionally, WDO embraced the United Nations Sustainable Development Goals (UN SDGs, also known as AGENDA 2030) as a call to action for its global community. Of particular relevance to LeNSes are inter alia: UN SDGs #7: Affordable and Clean Energy; #11: Sustainable Cities and Communities; #12: Responsible Consumption and Production; #13: Climate Change; and #17: Partnerships for the Goals.

A number of policies and strategies focusing on Africa have also been referenced. These include the Power Africa initiative, Agenda 2063 (of the African Union), as well as various national development plans. Consequently, the participation of African Higher Education Institutions (HEIs) as key catalysts involving local companies and practitioners is particularly encouraging. The knowledge co-generated in partnership with other international HEIs is a clear demonstration of the efficacy of international partnerships that seek to collaboratively solve some of the world's most urgent challenges. To this end, the twenty-first century offers a unique opportunity for Africa to leapfrog its human development and socioeconomic growth trajectories by tapping into the co-created didactic and pedagogic tools at their disposal. The easily accessible open-source and copyleft ethos adopted by the LeNSes initiative allows HEIs, as well as other participating entities within the so-called Quadruple Helix (of Academia; Business/Industry; Civil Society; and

Government) the opportunity to ideate and develop Product-Service System solutions to deliver context-responsive Distributed Renewable Energy systems, as well as to interrogate innovative case studies with respect to their unique priorities and resource capabilities. This will certainly make a significant contribution towards Africa's aspirations at producing well-informed future designers, engineers and allied professionals who are committed to sustainability and socioeconomic development in the broadest possible sense.

Cape Town, South Africa

Mugendi K. M'Rithaa
President Emeritus: World Design Organization

Acknowledgements

This volume is a collaboration of the following authors representing all partners in the LeNSes project, the Learning Network on Sustainable energy systems. A key editorial contribution to all of the chapters has been given by Carlo Vezzoli, Elisa Bacchetti, Fabrizio Ceschin, J. C. Diehl, Emanuela Delfino, Richie Moalosi and James Wafula.

Carlo Vezzoli[1] wrote: paragraphs 1, 1.1, 1.2.5, 1.3, 1.4, 1.5, 2.1, 2.2, 2.3, 4.1, 4.2, 4.2.1, 4.2.2, 4.2.3, 4.2.4, 4.3, 4.3.1, 4.3.2, 4.3.3, 5.1, 5.2, 5.3, 5.4, 5.6, 6.1, 6.2, 7.1, 7.2, 7.2.1, 7.2.2, 7.2.4, 7.2.8, 7.2.9, 7.2.10, 8.4; Chap. 3.

Mary Suzan Abbo[2] wrote: paragraphs 1.2.2.

Elisa Bacchetti[3] wrote: paragraphs 1.3, 2.2, 2.3, 2.3.1, 2.3.2, 2.3.3, 2.3.4, 2.3.5, 4.1, 4.2, 4.2.1, 4.2.2, 4.2.3, 4.2.4, 6.1, 6.2, 7.1, 7.2, 7.2.1, 7.2.2, 7.2.3, 7.2.4, 7.2.8, 7.2.9, 7.2.10, 8.1, 8.3.

Edurne Battista[4] wrote: paragraph 1.3.

Andrea Broom wrote: paragraph 1.2.3.

Fabrizio Ceschin[5] wrote: 2.2, 2.4, 4.3, 4.3.1, 4.3.2, 4.3.3, 4.4, 4.4.1, 4.4.2, 4.5, 4.5.1, 4.5.2, 4.5.3, 4.5.4, 4.5.5, 4.5.6, 5.1, 5.2, 5.3, 5.4, 5.5, 5.6. 7.2, 7.2.5, 7.2.6, 7.2.7, 8.1, 8.2.

Fiammetta Costa[6] wrote: paragraphs 1, 1.1, 1.2.5.

Emanuela Delfino[7] wrote: paragraphs 2.1, 2.2, 2.3, 2.3.1, 2.3.2, 2.3.3, 2.3.4, 2.3.5, 5.4, 6.1, 7.1, 7.2.

---

[1] Politecnico di Milano, Design Department, School of Design, Italy.

[2] Centre for Research in Energy and Energy Conservation (CREEC), College of Engineering, Design, Art and Technology (CEDAT) Makerere University, Uganda.

[3] Politecnico di Milano, Design Department, School of Design, Italy.

[4] Instituto de Investigación y Desarrollo Tecnológico para la Agricultura Familiar INTA-IPAF. Buenos Aires, Argentina.

[5] Brunel University London, College of Engineering, Design and Physical Sciences, Department of Design, Human Centred Design Institute, UK.

[6] Politecnico di Milano, Design Department, School of Design, Italy.

[7] Politecnico di Milano, Design Department, School of Design, Italy.

Jan Carel Diehl[8] wrote: paragraph 2.5.
Silvia Emili[9] wrote: paragraphs 2.2, 2.4, 4.3, 4.3.1, 4.3.2, 4.3.3, 4.4, 4.4.1, 4.4.2, 4.5, 4.5.1, 4.5.2, 4.5.3, 4.5.4, 4.5.5, 4.5.6, 7.2, 7.2.5, 7.2.6, 7.2.7, 8.1, 8.2.
Paulson Lethsolo[10] wrote: paragraph 5.7.
Richie Moalosi[11] wrote: paragraphs 1.2, 1.2.4, 5.7.
Mugendi K. M'Rithaa[12] wrote: paragraph 1.2.3.
Venny Nakazibwe[13] paragraph 1.2.2.
Mackay Okure[14] wrote: paragraph 1.2.2.
Lilac Osanjo[15] wrote: paragraph 1.2.1.
Yaone Rapitsenyane[16] wrote: paragraph 5.7.
Ephias Ruhode[17] wrote: paragraph 1.2.3.
James Wafula[18] wrote: paragraphs 1.2.1, 2.3.1, 2.3.2, 2.3.3, 2.3.4, 2.3.5.
All authors together wrote the notes from the book and paragraph 1.5.

[8]Delft University of Technology, Faculty of Industrial Design Engineering, The Netherlands.

[9]Brunel University London, College of Engineering, Design and Physical Sciences, Department of Design, UK.

[10]Department of Industrial and Technology, University of Botswana, Botswana.

[11]Department of Industrial and Technology, University of Botswana, Botswana.

[12]Cape Peninsula University of Technology, Industrial Design Department, South Africa.

[13]College of Engineering, Design, Art and Technology (CEDAT) Makerere University, Uganda.

[14]College of Engineering, Design, Art and Technology (CEDAT) Makerere University, Uganda.

[15]School of the Arts and Design (StAD), University of Nairobi, Kenya.

[16]Department of Industrial and Technology, University of Botswana, Botswana.

[17]Cape Peninsula University of Technology, Research, Innovation and Partnerships, Faculty of Informatics and Design, South Africa.

[18]Institute of Nuclear Science & Technology (INST), University of Nairobi, Kenya.

Notes from the book

This book reflects the main outcomes of the LeNSes (EduLink II programme, 2013–16 www.lenses.polimi.it) aimed to promote Design for Sustainability focused on sustainable energy access to all, as a crucial issue towards a sustainable society.

This book has beneficiated more widely from the contribution of several academics, researchers and designers from the LeNS worldwide network, which today includes more than hundred universities in five continents (www.lens-international.org). In particular from the LeNS Africa network (lensafrica.org.za), which currently involves 15 universities from the whole African continent, aiming to diffuse to designers, academics, professionals and students, the developed knowledge base and know-how on Design for Sustainability.

This book aims to share its contents with everyone who is interested to know more about designing Sustainable Product-Service System (S.PSS) applied to Distributed Renewable Energy (DRE), towards sustainable energy access for All.

Main contributions came from African and European partners of the LeNSes project, and particularly from Carlo Vezzoli (project coordinator), Fabrizio Ceschin, Lilac Osanjo, Mugendi K. M'Rithaa, Richie Moalosi, Venny Nakazibwe and Jan Carel Diehl, Elisa Bacchetti, Emanuela Delfino, Silvia Emili, Edurne Battista, Mackay Okure, Mary Suzan Abbo, Ephias Ruhode, Andrea Broom, James Wafula, Paulson Lethsolo and Yaone Rapitsenyane.

The book is organised to provide an overview of the topic and as well to support the design in practice. For this reason, the book includes **strategies** and **guidelines**, as well as a collection of **case studies** of Sustainable Product-Service System (S.PSS) applied to Distributed Renewable Energy (DRE) solutions. Additionally, are presented the method and support tools for designers.

The reading of this book can be supplemented by the **videos** and the **slides** of the lectures carried out during a set of pilot courses in the various LeNSes African partners universities; they are available on www.lenses.polimi.it, section courses. Coherently, the design **tools**, as well as the **case studies** and related **guidelines**, are accessible from the same website in an open and copyleft logic, i.e. available to be downloaded, adapted and reused in other contexts.

Contents

Part I Sustainable Energy for All

1 Energy and Sustainable Development 3
1.1 United Nations Sustainability Energy for All (SE4A) Agenda 4
1.2 Sustainable Energy for All in Africa 7
1.2.1 Sustainable Energy for All in Kenya 8
1.2.2 Sustainable Energy for All in Uganda 10
1.2.3 Sustainable Energy for All in South Africa 12
1.2.4 Sustainable Energy for All in Botswana 13
1.2.5 Sustainable Energy for All in Europe 15
1.3 Defining Access to Energy 16
1.4 Distributed Renewable Energy: A Key Leverage Towards Sustainable Energy for All (SE4A) 18
1.5 Sustainable Product-Service Systems Applied to Distributed Renewable Energy: An Introduction 19
References 21

2 Distributed/Decentralised Renewable Energy Systems 23
2.1 Distributed/Decentralised Renewable Energy: Sustainability 23
2.2 Distributed/Decentralised Renewable Energy Systems: Structures and Types 24
2.3 Renewable Energy Systems Types. 27
2.3.1 Solar Energy 28
2.3.2 Wind Energy 30
2.3.3 Hydro Energy 31
2.3.4 Biomass Energy 33
2.3.5 Geothermal Energy 34
2.4 Is Renewable Energy Zero Impact? 35
2.5 Barriers to Distributed/Decentralised Renewable Energy Systems 36
References 38

3 Sustainable Product-Service System (S.PSS) 41
3.1 S.PSS: An Introduction and Definition 41
3.2 S.PSS Types 42
3.2.1 Product-Oriented S.PSS: Adding Value to the Product Life Cycle (Type I) 42
3.2.2 Use-Oriented S.PSS: Offering Enabling Platforms for Customers (Type II) 43
3.2.3 Result-Oriented S.PSS: Offering Final Results to Customers (Type III) 44
3.3 S.PSS Sustainability Benefits 45
3.3.1 S.PSS Environmental Benefits 45
3.3.2 S.PSS Socioethical Benefits 47
3.3.3 S.PSS Economic and Competitive Benefits 48
3.4 S.PSS Barriers and Limits 48
3.4.1 Not All PSSs Are Sustainable 48
3.4.2 Barriers 48
References 49

4 Sustainable Product-Service System Applied to Distributed Renewable Energies 53
4.1 Sustainable Product-Service System Applied to Distributed Renewable Energy: A Win-Win Opportunity 53
4.2 Scenario for S.PSS Applied to Distributed Renewable Energy 54
4.2.1 Energy for All in Daily Life (Vision 1) 55
4.2.2 Energise Your Business Without Initial Investment Cost (Vision 2) 57
4.2.3 ‘Pay x Use’ Your Daily Life Products and Energy (Vision 3) 58
4.2.4 Start-up Your Business Paying Per Period for Equipment and Energy (Vision 4) 59
4.3 S.PSS Applied to DRE: Sustainability Potential Benefits 60
4.3.1 Environmental Benefits of S.PSS Applied to DRE 60
4.3.2 Socioethical Benefits of S.PSS Applied to DRE 61
4.3.3 Economic Benefits of S.PSS Applied to DRE 62
4.4 S.PSS Applied to DRE: A New Classification System and 15 Archetypal Models 62
4.4.1 Classification System 63
4.4.2 Archetypal Models of S.PSS Applied to DRE 68
4.5 S.PSS Applied to DRE: Critical Factors 83
4.5.1 Customer 84
4.5.2 Energy System 86
4.5.3 Services 87
4.5.4 Network of Providers 90
4.5.5 Offer 93
4.5.6 Payment Channels 95
References 97

Part II System Design for Sustainable Energy for All

5 Design for Sustainability: An Introduction 103
5.1 Evolution of Design for Sustainability 104
5.2 Product Life Cycle Design or Eco-Design 104
5.3 Design for Eco-Efficient Product-Service Systems 105
5.4 Design for Social Equity and Cohesion 106
5.5 Design for Socio-Technical Transitions 107
5.6 State of the Art of Design for Sustainability 107
5.7 Human-Centred and Universal Design 108
References 118

6 System Design For Sustainable Energy For All: A New Role For Designers 125
6.1 System Design for Sustainable Energy for All 125
6.2 SE4A Design Criteria, Guidelines and Examples 126

Part III Method and Tools for SD4SEA

7 Method and Tools for System Design for Sustainable Energy for All 141
7.1 Method for System Design for Sustainable Energy for All 141
7.2 SD4SEA Tools 151
7.2.1 Sustainability Design Orienting Scenario (SDOS) on S.PSS&DRE 151
7.2.2 Strategic Analysis (SA) Template 154
7.2.3 Sustainable Energy for All Idea Tables and Cards 155
7.2.4 E.DRE—Estimator for Distributed Renewable Energy 159
7.2.5 PSS + DRE Innovation Map 166
7.2.6 S.PSS + DRE Design Framework & Cards 173
7.2.7 The Energy System Map 182
7.2.8 Innovation Diagram for S.PSS&DRE 189
7.2.9 Concept Description Form for S.PSS and DRE 194
7.2.10 Stakeholder Motivation and Sustainability Table 195
References 197

8 Practical Examples of Application of SD4SEA Approach/Tools 199
8.1 Introduction 199
8.2 Solar Energy Company, Botswana 200
8.3 SMEs for Energy, Uganda 204
8.4 Summary and Considerations 207

List of Figures

Fig. 2.1 Paradigm shift from non-renewable/centralised energy generation systems to renewable/distributed ones. *Source* designed by the Authors . . . 24
Fig. 2.2 Centralised energy system. *Source* designed by the Authors. . . . 25
Fig. 2.3 Decentralised energy system. *Source* designed by the Authors . . . 26
Fig. 2.4 Distributed energy system. *Source* designed by the Authors. . . . 26
Fig. 2.5 Distributed/decentralised energy. System structure and configurations. *Source* designed by the Authors . . . 27
Fig. 2.6 World map solar horizontal irradiation. *Source* https://solargis.com/legal/terms-of-use-for-ghi-free-maps/ . . . 28
Fig. 2.7 Solar heaters components. *Source* www.ashden.org . . . 30
Fig. 2.8 Process of gasification. *Source* www.ashden.org. . . . 34
Fig. 3.1 S.PSS: a paradigm shift from traditional product offer. *Source* designed by the Authors . . . 42
Fig. 4.1 The coupling of the 2 paradigm shifts represented by S.PSS and DRE. *Source* designed by the Authors. . . . 55
Fig. 4.2 The sustainable energy for all scenario. *Source* designed by the Authors . . . 55
Fig. 4.3 Screenshots from the video 'Energy for all in daily life'. *Source* Vanitkoopalangkul [58]. . . . 56
Fig. 4.4 Screenshots from the video energise your business without investment cost. *Source* Vanitkoopalangkul [58] . . . 57
Fig. 4.5 Screenshots from the video 'Pay x use' for your daily life products and energy. *Source* Vanitkoopalangkul [58]. . . . 58
Fig. 4.6 Screenshots from the video 'Start-up your Business' paying per period for equipment and energy. *Source* Vanitkoopalangkul [58] . . . 60
Fig. 4.7 Selection of dimensions' polarities and combination of axis used to build the classification system. *Source* Emili et al. [17] . . . 65

Fig. 4.8 Classification system. *Source* Emili [18]. 67
Fig. 4.9 Classification system with archetypal models. *Source* Emili [18] 69
Fig. 4.10 Archetypal model 1: selling individual energy systems with advice and training services. *Source* Emili [18] 70
Fig. 4.11 Archetypal model 2: offering advice and training services for community-owned and community-managed isolated mini-grids. *Source* Emili [18] 70
Fig. 4.12 Archetypal model 3: offering advice and training services for community-owned and community-managed connected mini-grids. *Source* Emili [18] 71
Fig. 4.13 Archetypal model 4: selling mini-kits with additional services. *Source* Emili [18] 72
Fig. 4.14 Archetypal model 5: selling individual energy systems with additional services. *Source* Emili [18]. 73
Fig. 4.15 Archetypal model 6: offering individual energy systems (and Energy-Using Products) in leasing. *Source* Emili [18] 74
Fig. 4.16 Archetypal model 7: renting Energy-Using Products through entrepreneur-owned and entrepreneur-managed charging stations. *Source* Emili [18] 75
Fig. 4.17 Archetypal model 8: renting Energy-Using Products through entrepreneur or community-managed charging stations. *Source* Emili [18] 76
Fig. 4.18 Archetypal model 9: offering access to energy (and Energy-Using Products) on a pay-per-consumption basis through individual energy systems. *Source* Emili [18] 77
Fig. 4.19 Archetypal model 10: offering access to energy (and Energy-Using Products) on a pay-per-consumption basis through isolated mini-grids. *Source* Emili [18] 78
Fig. 4.20 Archetypal model 11: offering access to energy and Energy-Using Products on a pay-per-unit of satisfaction basis through mini-kits. *Source* Emili [18] 79
Fig. 4.21 Archetypal model 12: offering access to energy (and Energy-Using Products) on a pay-per-unit of satisfaction basis through individual energy systems. *Source* Emili [18]. . . . 80
Fig. 4.22 Archetypal model 13: offering access to energy (and Energy-Using Products) on a pay-per-unit of satisfaction basis through individual energy systems. *Source* Emili [18] 81

Fig. 4.23 Archetypal model 14: offering access to Energy-Using Products through community- or entrepreneur-managed charging stations on a pay-per-unit of satisfaction basis. *Source* Emili [18] . . . 81
Fig. 4.24 Archetypal model 15: offering access to energy (and Energy-Using Products) on a pay-per-unit of satisfaction basis through mini-grids. *Source* Emili [18] . . . 82
Fig. 5.1 Human-centred design pyramid. *Source* Giacomin [42] . . . 112
Fig. 6.1 Visualization of the System Map for an 'Appropriate Stakeholder Configuration Design'. *Source* designed by the Authors . . . 126
Fig. 6.2 Indigo, Sub-Saharan Africa. *Source* www.azuri-technologies.com. . . . 127
Fig. 6.3 Bboxx, Africa and Asia. *Source* www.bboxx.co.uk . . . 130
Fig. 6.4 Sunlabob, Laos. *Source* www.sunlabob.com. . . . 131
Fig. 6.5 Husk Power Systems, India. *Source* www.huskpowersystems.com . . . 134
Fig. 6.6 Solar-Powered Café, South Africa. *Source* www.kutengatechnology.com . . . 136
Fig. 6.7 Domestic Biogas, Nepal. *Source* www.ashden.org . . . 137
Fig. 7.1 Sustainability design orienting scenario on S.PSS and DRE. *Source* designed by the Authors. . . . 153
Fig. 7.2 Strategic analysis (SA) template. *Source* designed by the Authors . . . 154
Fig. 7.3 Printable sustainable energy for all idea tables. *Source* designed by the Authors . . . 156
Fig. 7.4 Cards (with case studies) to complement the sustainable energy for all idea tables. *Source* designed by the Authors. . . . 157
Fig. 7.5 Digital version of sustainable energy for all idea tables and cards (with case studies). *Source* designed by the Authors . . . 158
Fig. 7.6 S.PSS and DRE estimator of DRE load/need and production potential. *Source* designed by the Authors . . . 160
Fig. 7.7 Worksheet—solar energy. *Source* designed by the Authors . . . 162
Fig. 7.8 Worksheet—wind energy. *Source* designed by the Authors . . . 163
Fig. 7.9 Worksheet—hydro energy. *Source* designed by the Authors . . . 164
Fig. 7.10 Worksheet—biomass energy. *Source* designed by the Authors . . . 165
Fig. 7.11 PSS + DRE innovation map. *Source* Emili [7]. . . . 167
Fig. 7.12 Archetypes cards. *Source* [7] . . . 168
Fig. 7.13 Stakeholders card. *Source* Emili [7] . . . 168
Fig. 7.14 Concept card. *Source* Emili [7]. . . . 169

Fig. 7.15 Positioning of company's offerings on the innovation map. *Source* designed by the Authors. 170
Fig. 7.16 Positioning of competitors on the innovation map. *Source* designed by the Authors 170
Fig. 7.17 Example of a completed stakeholder card. *Source* designed by the Authors 171
Fig. 7.18 Selection of promising areas to explore. *Source* designed by the Authors 172
Fig. 7.19 Example of a completed concept card. *Source* designed by the Authors 172
Fig. 7.20 Example of a completed innovation map. *Source* designed by the Authors 173
Fig. 7.21 The design framework, design canvas and cards. *Source* designed by the Authors 175
Fig. 7.22 Design framework. *Source* Emili [7] 176
Fig. 7.23 List of cards for each design element. *Source* Emili [7] 177
Fig. 7.24 Example of cards' structure. *Source* Emili [7] 178
Fig. 7.25 The design canvas. *Source* Emili [7] 179
Fig. 7.26 Positioning ideas on the design canvas. *Source* designed by the Authors 181
Fig. 7.27 The energy system map tool: icons, example and template. *Source* designed by the Authors 184
Fig. 7.28 Example of energy system map. *Source* designed by the Authors 185
Fig. 7.29 Icons. *Source* Emili [7]. 186
Fig. 7.30 Legend for the energy system map. *Source* Emili [7]. 187
Fig. 7.31 Positioning icons on the template. *Source* designed by the Authors 188
Fig. 7.32 Innovation diagram for S.PSS&DRE. *Source* designed by the Authors 190
Fig. 7.33 Labels to support the innovation diagram for S.PSS and DRE. *Source* designed by the Authors 190
Fig. 7.34 Concept description form for S.PSS and DRE. *Source* designed by the Authors 194
Fig. 7.35 Stakeholders motivation and sustainability table. *Source* designed by the Authors 196
Fig. 8.1 Innovation map completed by the SME in Botswana. *Source* designed by the Authors 201
Fig. 8.2 The design canvas produced by the company. *Source* designed by the Authors 203
Fig. 8.3 The energy system map produced by the company. *Source* designed by the Authors 204
Fig. 8.4 Current business of a SME involved in the course. *Source* designed by the Authors 205

Fig. 8.5 Screenshot from sustainability design orienting scenario for S.PSS and DRE. *Source* designed by the Authors 206
Fig. 8.6 Ideas generated using the SE4All idea generation tables and cards. *Source* designed by the Authors 206
Fig. 8.7 Stakeholders' motivation and sustainability table generated by participants. *Source* designed by the Authors 207

List of Tables

Table 1.1 SE4All initiative Kenya targets 9
Table 2.1 Classification of barriers to the diffusion of DRE. 38
Table 4.1 List of S.PSS and DRE dimensions 63
Table 4.2 Dimensions characterising S.PSS applied to DRE 64
Table 5.1 Human-centred design tools 116
Table 7.1 Stages, aims, processes and tools for each stage of the method for SD4SEA 142

Part I
Sustainable Energy for All

Chapter 1
Energy and Sustainable Development

The world is facing a strong evolution due to the advancement of information and communication technologies that set the knowledge technologies at the base of productivity, competition and power. The world is more and more interconnected than ever before, i.e. people, ideas, images, goods and money are being distributed more frequent and faster than ever before. We live in a network society, which is not divided into independent and isolated nations or communities, and at the same time enterprises are organised in network, i.e. there has been an increase of teamwork, networking, outsourcing, subcontracting and delocalisation. All these features may represent the advancement of our civilization, but at what price are we paying for the environmental and socioethical impacts?

Historically, we have discovered that the production and consumption system did not only produce advantages, but also disadvantages. This happened in the economic boom of the 1960s when industrialised countries faced a strong acceleration of consumption and production system development.

Since that moment, we became aware that human activities may determine harmful and irreversible environmental impacts, and it carries the notion of environmental limits.

It was in 1972 when the book *Limits to Growth* [12] was published based on a first computerised simulation of the effects on the nature of the ongoing system of production and consumption. It was the first scientific forecast of a possible global eco-system collapse. Fifteen years later, in 1987, the United Nations World Commission for Environment and Development (WCED) provided the first definition of Sustainable Development:

> A social and productive development that takes place within the limits set by the "nature" and meets the needs of the present without compromising those of the future generation within a worldwide equitable redistribution of resources.

In fact, this incorporates even the fundamental challenge of social equity and cohesion (i.e. the socioethical dimension of sustainability).

C. Vezzoli et al., *Designing Sustainable Energy for All*,
Green Energy and Technology, https://doi.org/10.1007/978-3-319-70223-0_1

In the recent period, the concept of sustainable development has been linked to the one of accesses to sustainable energy. Indeed, it has become a shared understanding that sustainable development is not possible without sustainable energy access to all. Energy is the world's largest industrial sector ($\sim$70% of world GDP) whose output is an essential input to almost every good and service provided in the current economy. Energy services have a profound effect on productivity, health, education, food and water security, and communication services. Therefore, that access to energy can contribute to reduce inequality and poverty.

Very often, problems of the production system are only related to materials impacts, i.e. residues, pollution caused by cars, planned obsolescence: which we can see and experiment the effects of it. However, energy represents the hidden side of others. First, energy enables us to produce things by the way we do, and environmental impacts start with the transformation of a given resource.

On the other hand, there are implications connected to energy. Transforming resources into energy requires the capability (in terms of technologies) and the financial resources to face it. At the same time, our current energy system, based on a fossil fuel model, implies a kind of resource that is not available in all the countries. Both features—localization and budget—mean that there is interest around energy, which include politics and economics issues. What is clear for now is that only those who have the control of the energy system have the possibility to increase their development. Access or no access to energy determines our quality of life and its limited access represent one of the key barriers to achieve sustainable development.

1.1 United Nations Sustainability Energy for All (SE4A) Agenda

Sustainable development emerged as a major global issue back in the 1970 with the publication of the report 'Limits to Growth' [12]. In the 1980 and 1990 milestones such as the Brundtland Report (Our Common Future) by the United Nations World Commission for Environment and Development [22] and the Earth Summit held in Rio de Janeiro in 1992 paved the way to worldwide acknowledgement for the necessity of major changes related to environmental and social pressures now felt as a global problem. Not only it gained public recognition but achieved a stage of maturation, with new policies being created and implemented at various scales.

More recently, the United Nations General Assembly designated the year 2012 as the International Year of Sustainable Energy for All and unanimously declared 2014–2024 as the Decade of Sustainable Energy for All. *United Nations Secretary-General* Ban Ki Moon has appointed a High-Level Group on the same topic, which delivered a Global Action Agenda prior to the UN Conference on Sustainable Development (Rio + 20). As Ban Ki Moon stated launching the Sustainable Energy for All Initiative [21], '*Energy poverty is a threat to the achievement of the Millennium Development Goals. At the same time, we must move very rapidly toward a clean energy economy to prevent the dangerous warming of our planet*'.

The Sustainable Energy for All Initiative, identified three inter-linked objectives to be achieved by 2030 and pursued during the SE4All decade, necessary for long-term sustainable development in relation to access to energy:

- ensure universal access to modern energy services;
- double the rate of improvement in energy efficiency;
- double the share of renewable energy in the global energy mix.

To continue pursuing the above efforts, expressed by the Sustainable Energy for All Initiative, the Sustainable Development Goal number 7 of the Global Action Agenda [20] advocates for the need to ensure access to affordable, reliable, sustainable and modern energy for all.

The SDG number 7 targets that by 2030 the following should have been achieved:

- Ensure universal access to affordable, reliable and modern energy services;
- Increase substantially the share of renewable energy in the global energy mix;
- Double the global rate of improvement in energy efficiency;
- Enhance international cooperation to facilitate access to clean energy research and technology, including renewable energy, energy efficiency and advanced and cleaner fossil fuel technology, and promote investment in energy infrastructure and clean energy technology;
- Expand infrastructure and upgrade technology for supplying modern and sustainable energy services for all in developing countries, least developed countries, small island developing states and land-locked developing countries, in accordance with their respective programmes of support.

A study conducted by Rogelj et al. [16] on the compatibility of the 'Sustainable Energy for All' initiative with a warming limit of 2 °C shows that achieving the three energy objectives could provide an important entry point to climate protection, and that sustainability and poverty eradication can go hand in hand with mitigating climate risks. However, the researchers warn that the likelihood of reaching climate targets within the scenarios depends as well on a variety of other factors, including future energy demand growth, economic growth and technological innovation. Therefore, securing energy for all within the existing environmental boundaries requires further political measures and financial resources. According to Nilsson [14] '*Investment costs for these pathways are large but often profitable for society and most of them have already been set in motion. Still, progress is slow and must be accelerated at national and regional levels. Carbon pricing is necessary but not sufficient: beyond this, governance responses need to be put in place to induce transitions through scaling up a diversity of supply and demand options. White and green certificates, feed-in tariffs, technology standards and removal of fossil subsidies are important first steps already under way. These contribute to nurturing and scaling up new technological regimes, as well as destabilizing old and unsustainable ones*'.

The Sustainable Energy for All Global Action Agenda defines specific requirements for different contexts. Low- and middle-income country governments must create conditions that enable growth by establishing a clear vision, national targets, policies, regulations and incentives that link energy to overall development, while strengthening national utilities. More than 80 governments from low- and middle-income countries have joined the SE4A initiative. Industrialised country governments must focus internally on efficiency and renewable energies while externally supporting all three objectives through international action. They elaborate on current plans to increase the deployment of domestic renewable energy and improve energy efficiency through the entire value chain, from production of primary energy-using energy services. The Global Action Agenda highlights also sectoral action areas addressing both power generation and the principal sectors of energy consumption. These include

- Modern cooking appliances and fuels;
- Distributed electricity solutions;
- Grid infrastructure and supply efficiency;
- Large-scale renewable power;
- Industrial and agricultural processes;
- Transportation;
- Buildings and appliances.

It is important to underline that the sectoral actions have to be combined in order to assure immediate basic energy access to improved quality of life and well-being, but also to build energy services for long-term autonomous sustainable development.

According to the Agenda, those solutions include all distributed options for electrification, which range from island-scale grid infrastructure to mini-grids to much smaller off-grid decentralised individual household systems and targeted applications for productive uses. Experience has demonstrated that the best progress has come in low- and middle-income contexts that pursued strategies and policies to expand access to all (i.e. both urban and rural communities) by including the full range of electrification options in a balanced way. The *World Energy Outlook* 2011 [8] concludes that grid extension is the best option for achieving universal access in all urban areas but in only 30% of rural areas. The IEA projects [8] that around 45% of the additional connections needed for universal access will come from grid expansion, while the remaining 55% will depend on micro-grids and off-grid solutions.

In distributed electricity solutions, opportunities can be perceived for the involvement of different stakeholders, i.e. governments, donors, businesses and civil society.

Examples of already active initiatives that fall into this area are Lighting Africa and Lighting Asia, driven by the World Bank and International Finance Corporation (IFC); Lighting a Billion Lives under The Energy and Resources Institute (TERI); regional development banks' distributed energy projects such as those promoted under 'Energy for All' by Asian Development Bank (ADB) and by African Development Bank (AfDB) under the Scaling-Up Renewable Energy Programme in Low-Income Countries; African Caribbean Pacific, Europe (ACP-EU) Energy

Facility-Energy Project of United Nations Development Programme and Global Environment Facility (UNDP/GEF); and Global Lighting and Energy Access Partnership (LEAP) led by U.S. Department of Energy.

1.2 Sustainable Energy for All in Africa

Africa is the second largest continent, with over 2000 languages spoken in the 54 nations. The burgeoning youthful population and abundance of human and natural resources inspire optimism for unprecedented growth as we advance into the twentieth century. Additionally, 2008 marked the first time in human history when more people lived in urban areas than rural one—a phenomenon that has a far more dramatic impact on developing regions of the world such as those found in Africa. Instructively, since 2011, six of the twelve fastest growing economies are from the African continent. This increased socioeconomic development has led to greater demand for food, shelter and energy (among other key resources).

In Africa, the speed at which distributed and networked technologies are proliferating is quite interesting. Examples abound from mobile telephony and ICT of cost-effective and accessible product-service-system offerings that make the continent an ideal context for the deployment of distributed solutions. Whereas the continent abounds with minerals and myriad natural resources, the majority of its denizens still do not have access to adequate housing, water, electricity and related basic needs to help propel its communities into a truly sustainable future. To this end, sustainable energy systems are crucial and indispensable to desired socioeconomic development. Further, the massive size of the continent demands creative distributed systems that take cognisance of the sociotechnical and geopolitical aspirations of myriad societies.

There is a school of thought that Africa will be unable to alleviate poverty and improve the well-being of its people, reduce inequalities, if it cannot sustainably produce its own energy. Africa has abundant sunshine and vast water resources which can be used to generate cleaner, cheaper and accessible sustainable energy. On the contrary, over 600 million people in Africa still live in darkness without electricity. This lack of access to electricity has reduced the continent's economic growth, quality of education especially in rural areas and greatly affected health facilities and agricultural activities. It is not yet late to reverse this challenging scenario. This challenge provides an opportunity to critically think about clean, efficient, resilient and low-carbon technologies and sustainable development to reduce overdependence on fossil fuels. Access to sustainable energy will cut household costs, releasing resources to productive health and education investment as well as boosting the renewable energy businesses. This has the potential to drive economic growth and create jobs. In 2011, the United Nations launched the Sustainable Energy for All (SE4All) initiative to ensure universal access to modern energy services, doubling the global rate of improvement in energy efficiency, and doubling the share of renewable energy in the global mix. The aim is to achieve these three goals by 2030.

Such an initiative provides Higher Education Institutions (HEIs) in the continent with a unique opportunity to contribute to efforts at capacitation, research and pedagogy in redressing the pressing challenges associated with the quest for sustainable energy security. Notwithstanding, dedicated research, design and development initiatives focusing on sustainable energy systems are few and far between.

1.2.1 Sustainable Energy for All in Kenya

Kenya opted to be part of the SE4All UN Initiative because the Government had achieved significant strides in developing the framework for energy development, thanks to the Energy Policy, 2004, and Energy Act, 2006. Review of these two documents is expected to further improve the enabling environment for the engagement of a wide range of stakeholders, and particularly private sector, in the delivery of clean and modern energy services. It also happens at a time when petroleum resources have been discovered in the country and will therefore be instrumental in diversifying the energy mix and addressing energy poverty.

The SE4All Action Agenda (AA) for Kenya[1] presents an energy sector-wide long-term vision spanning the period 2015–2030. It outlines how Kenya will achieve her SE4All goals of 100% universal access to modern energy services, increase the rate of energy efficiency and increase to 80% the share of renewable energy in her energy mix, by 2030 (Table 1.1).

Biomass

In the context of the SE4All, access to modern energy involves electricity and energy for cooking. Kenya has chosen the baseline year for electricity access as 2012. For the purpose of the AA, the definition of electricity access is connections to the national grid system or distributed (off-grid) electricity solutions which include Solar Home Systems (SHS) and mini-grids. In the baseline year, only 23% of the population, which represents 1.97 million households, had grid electricity supply. Access to modern cooking services refers to access to improved cookstoves and non-solid fuels. The baseline year for access to improved cookstoves was 2013, being at the level of 3.2 million households, according to market assessment of Clean Cookstoves Association of Kenya (CCAK) under the Kenya Country Action Plan 2013 (KCAP). Over 80% of Kenyans rely on the traditional use of biomass as the primary source of energy for cooking and heating, with firewood contributing 68.7% and charcoal 13.3%. The Kenyan government is putting in place measures to regulate the fuelwood sector with a draft Forest Act[2] envisaging a six-point system of control from producer to consumer.

---

[1]SE4All > Action Agenda for Kenya: www.se4all.org/sites/default/files/Kenya_AA_EN_Released.pdf.

[2]The 2009 charcoal production regulations developed by the Kenya Forest Service are yet to be adopted.

Table 1.1 SE4All initiative Kenya targets

Universal access to modern energy services	Doubling global rate of improvement of energy efficiency	Doubling share of renewable energy in global energy mix	Universal access to modern energy services	
Percentage of population with electricity access	Percentage of population with access to modern cooking solutions	Rate of improvement in energy intensity	Renewable energy share in Total Final Energy Consumption	
			Power	Heat
100%[a]	100%	−2785%/year[b]	80%	80%

Legend [a]Projected to be reached by 2022
[b]The energy intensity is expressed in negative as its improvement is a reduction on the energy intensity
Source Beyond Connections: Energy Access Redefined, Technical Report, Energy Sector Management Assistance Program, World Bank Group and SEforALL

Biomass contribution to Kenya's final energy demand is 69% and provides for more than 90% of rural household energy needs. The main sources of biomass for Kenya include charcoal, wood fuel and agricultural waste.[3] Fuelwood demand is at 35 million tonne per year, while the supply is at 15 million tonne per year representing a deficit of 20 million tonne. The deficit is largely the cause of high rate of deforestation, resulting in adverse environmental effects such as desertification, land degradation, drought and famine. One of the ways of arresting this is through the promotion of improved cooking stoves.

Because rural energy suffers low priority and status in both planning and development resource allocation, the Energy Bill 2015 proposes the establishment of the Rural Electrification and Renewable Energy Corporation. Amongst other functions, the Corporation will develop and update the renewable energy master plan taking into account County-specific needs and the principle of equity in the development of renewable energy resources. The Bill also proposes the establishment of energy centres in the Counties and a framework for collaboration with the County Governments in the discharge of its mandate. This framework includes undertaking on-farm and on-station demonstration of wood fuel species, seedling production and management in order to address the deficit in the national fuelwood demand.

Electricity

According to the March 2011 Least Cost Power Development Plan (2011–2031),[4] the required installed capacity for the reference scenario in 2030 will be 15,065 MW. The present value for this installed capacity amounts to €34.8 billion, (committed projects excluded) expressed in constant prices as of the beginning of 2010.

[3]Source www.erc.go.ke, 2016.

[4]Complete information about Least Cost Power Development Plan (2011–2031), available at http://www.renewableenergy.go.ke/index.php/content/44.

The transmission development plan indicates the need to develop approximately 10,345 km of new lines at an estimated present cost of €3.8 billion. Transmission development during the planning horizon will be based on 132,220 and 400 kV. According to the 5-year (2013–2017) corporate strategic plan for the electricity sub-sector, Kenya targets installed capacity of 6762 MW consisting of 49.9% Renewable Energy, 15.5% Natural Gas, 28.4% Coal and 6.2% diesel by 2018. The total generation capital expansion cost up to 2018 cost is estimated at €6.5 billion under the moderate estimations.

There are 41 transmission investment programmes associated with implementation of the additional 5000+MW investment by 2018 at an estimated cost of €3.1 billion. The corporate strategy plan targets 3325 km of new transmission lines and 3178 MVA of new transmission substation capacity for transmission systems and 3768 km of new MV lines.

The distribution system targets 69 new substations of capacity 6225 MVA; 20 new bulk supply points of capacity 1237.5 MVA for distribution systems and 70% household connectivity to electricity. The estimated cost of implementing the distribution system is €1.1 billion.

Implementers of transmission and distribution projects are Kenya Electricity Transmission Company (KETRACO) and Kenya Power and Lighting Company (KPLC), respectively.

1.2.2 Sustainable Energy for All in Uganda

Electrification access in Uganda stands at approximately 26.1% nationally (14.88% centralised grid and 11.22% decentralised) and 7% in rural areas. Since 2001, the government of Uganda has stepped up its efforts to extend energy access to the rural communities. Several statutory agencies (Central Government, Local Governments, civil society, the private sector and international agencies) are key contributors to the institutional framework for energy access. The Ministry of Energy and Mineral Development is the lead Government body responsible for policy development, guidance and implementation in the energy sector. Its activities are grounded in the national development plan.

The National Development Plan foresees investment in the energy infrastructure to raise electricity consumption from 75 to 674 kWh/capita, a rate comparable to that of Malaysia and Korea. Hence, generation capacity will be increased to meet the needed 3500 MW. Work has started for the construction of several hydropower production plants, namely, Bujagali HPP 250 MW, Karuma HPP 700 MW, Ayago HPP 700 MW, Isimba HPP 130 MW and Arianga HPP 400 MW. It is also envisaged that additional energy shall be generated from renewable sources as follows: Thermal plants 700 MW, Mini HPP 150 MW, Solar thermal 150 MW, Geothermal 150 MW and cogeneration from biomass 150 MW. Consequently, rural electrification, which currently stands at 4%, is expected to increase by 20% and reduction of power losses by 16%.

The Rural Electrification Agency (REA) was established as a semi-autonomous agency by the Minister of Energy and Mineral Development through Statutory Instrument 2001 No. 75, to operationalise Government's rural electrification programme. During the same year, having observed that the forest cover in Uganda is fast diminishing, the shrinking rate being estimated at 55,000 ha per year or 2%, a Forestry Policy was passed. The Forestry Policy assigns the responsibility of developing and implementing strategies for biomass energy conservation, focusing on households, charcoal producers and industrial consumers to the MEMD.

Subsequently, in 2002, the government passed the Uganda Energy Policy, and in 2007 the Renewable Energy Policy was enacted. The overall objective of the Renewable Energy Policy is to diversify the energy supply sources and technologies in the country. In particular, the Policy goal strives to increase the use of modern renewable energy from the current 4–61% of the total energy consumption by the year 2017. The operationalization of the Renewable Energy Policy culminated in the establishment of a Renewable Energy Department and an Energy Efficiency and Conservation Department in the Ministry of Energy and Mineral Development, establishing a National Energy Committee at the National Level and District Energy Committees and District Energy Offices at the Local Governments.

With the above policies in place, and with support from the development partners, the promotion of sustainable energy resources has received significant attention. Currently, three factories in Uganda, namely, Kakira Sugar Works Ltd., Kinyara Sugar Works Ltd. and Sugar Corporation of Uganda Ltd.—run cogeneration plants based on bagasse. The total capacity is 22 MW. Out of this, 12 MW from Kakira Sugar Works is supplied to the grid. Several industries have also embraced the use of wood chips from the carpentry and coffee husks as alternative sources of energy. The use of improved stoves is currently promoted by the Ministry of Energy and Mineral Development with support of the Uganda German Development Corporation through the Promotion of Renewable Energy and Energy Efficiency Programme (PREEEP).

In addition, the government of Uganda through its rural electrification programme is promoting the use of solar energy in the areas that have no access to the grid. This programme also involves extension of low and medium voltage lines in the rural areas. So far, over 3000 km of Medium Voltage lines (33 and 11 kV) and 2500 km of Low Voltage lines have been constructed and commissioned and an additional 2100 km of MV and 1000 km of LV are currently under construction. A total of 1280 rural communities (villages, trading centres, social centres and public institutions) with a potential of 120,000 connections have access to electricity and at least 38,530 connections have been achieved outside the main grid (Development of Indicative Rural Electrification Master Plan—2009). Two private companies and two cooperatives were awarded operation and maintenance concessions in seven areas of the country for large regional lines outside UMEME areas of operation. Ready boards have been introduced to ease connection of poorer households to electricity. Besides, in order to streamline consumption and payment of bills, the use of prepaid metering has been introduced and the Rural Electrification Agency (REA) has awarded concessions to users in rural areas.

1.2.3 Sustainable Energy for All in South Africa

National Energy Efficiency Strategy
The draft National Energy Efficiency Strategy under the auspices of the Department of Energy is currently undergoing revision.

Sustainable Energy Strategy for the Western Cape
A recent energy crisis in the Western Cape has highlighted the need to develop a plan for sustainable, secure energy provision in the Western Cape. Although various national efforts are underway to increase energy provision to the Western Cape, the Provincial Government believes that additional efforts need to be made to address the other energy challenges facing the Province, including the challenges of

- Reducing the Province's carbon footprint;
- Providing access to energy to all citizens in the province, and
- Addressing the numerous health, social and environmental problems associated with our current energy use patterns.

These challenges need to be addressed in the context of supporting the Province's economic development and job creation. The development of this discussion document was preceded by a Status Quo and Gap Analysis which highlighted the need for an effective energy policy to ensure the availability of background information and data for policy-makers, provide an effective institutional structure for sustainable energy management, develop a regulatory and policy framework, develop a training, communications and awareness raising programme and establish partnerships with public and private sector bodies.

Based on the gaps identified, certain actions have already been taken (Western Cape Government 2007), including

- The formation of an Intergovernmental Energy Task Team (IETT);
- Ongoing engagement with stakeholders at provincial and national level;
- Completion of a provincial energy inventory, which has been used to inform the adoption of a resolution at the Sustainable Development Conference requiring the Province to develop a strategy to address energy and climate change.

Skills Development for the Green Economy
The vision of the Western Cape Government (WCG) *Skills Development for the Green Economy* (2013) is a knowledge-driven project being championed by the CHEC-WCG Coordinating Group on Climate Change:

> The future of the South African economy is threatened by poverty and unemployment, the impact of climate change, declining and degraded natural resources. Solving these problems lies in a transition to a green economy, one characterised by low carbon emissions, the efficient use of resources and social inclusion (2013:3).

The vision of the Western Cape Government is to be the centre of this transition in South Africa, to make the Province a 'Green Economic Hub' for green investment and business opportunities that alleviate poverty, restore degraded eco-systems that provide essential services to society, and achieve energy, water and food security. To realise this vision, the Provincial Government has produced a Green Economy Strategy to outline a framework for public, private and community sectors to co-operatively pursue this green economic growth. The Green Economy Strategy itself is informed by and arose from the requirements of the Western Cape Climate Change Response Strategy, which highlights the need for planning, preparation and innovation to maximise the province's capacity to adapt to the impacts of climate change. However, there is currently a lack of suitably qualified professionals and technicians to successfully implement the Climate Change Response Strategy and Green Economy Strategy. The Province has thus identified an urgent need for skills development in the areas of climate change mitigation and adaptation, the green economy and infrastructure development. To this end, representatives of the Western Cape Government (WCG) met with the Cape Higher Education Consortium's (CHEC) Coordinating Group on Climate Change, and communicated the need to match university education with the knowledge and core competencies that will be required by the Climate Change Response Strategy and Green Economy Strategy. Subsequently, the task team was requested to extend the scope of the project to cover Further Education and Training Colleges if support from the CEOs of the colleges was provided and suitable researchers could be found.

The vision of the CHEC representatives is for the four universities to collaborate with one another and with the Province to provide up-to-date, relevant education in the areas of the green economy and climate change, and equip professionals with the skills and core competencies necessary to bring the 'Green Economic Hub' to life.

1.2.4 Sustainable Energy for All in Botswana

As of 2013, the total population in Botswana which had access to electricity was 66% and of these, 75% were in urban areas and 54% in rural areas (Statistics Botswana 2017). About 700,000 people do not have access to electricity, mainly those who are far away from the main grid. In Botswana, 98.5% of the electricity is generated from fossil fuels and only 1.5% is generated from renewable sources (Statistics Botswana 2017). These statistics show that Botswana is still far from achieving sustainable energy for all due to over-reliance of generating electricity from fossil fuels.

Botswana Power Corporation is the sole parastatal utility which was formed in 1970 by an Act of Parliament and its mandate is to generate, transmit and distribute electricity within Botswana. According to the Corporation's mandate, throughout the years, efforts were focused on reducing the Corporation's activities' impact on wildlife, Greenhouse Gas (GHG's) emissions and the landscape and thus striking a balance between the interests of industry and the effective use and conservation of resources.

The Corporation has initiated the replacement of insulated low voltage overhead line conductor with insulated Ariel Bundled Conductor (ABC) in the distribution network and this blends in well with the above environmental concerns, in that little if any tree clearing is done to facilitate low voltage line construction.

Generation waste management by the Corporation has focused on the following activities:

- Ash disposal;
- Waste water treatment;
- Groundwater pollution monitoring;
- Greenhouse gas emission monitoring.

As a mitigation measure, the Corporation supports research projects for provision of electricity using other efficient alternatives to thermally generated electricity and is working with some international entities that are currently involved in a pilot project for the supply of electricity using alternative sources of energy such as solar energy. As a project to encourage efficient use of energy, Botswana Power Corporation has installed 1 million Compact Fluorescent Lamps (CFLs) in households across the country and this will greatly save energy. To address some of the challenges, an environmental or a sustainable development policy is being formulated that will serve the Corporation and the nation well into the future.

Botswana Power Corporation has a 1.3 MW Photovoltaic solar power plant at Phakalane, a suburb of Gaborone. The solar power project was implemented through a Japanese grant as part of a strategy called '*Cool Earth Partnership*' which Japan announced in 2008 to address environmental issues. Under the *Cool Earth Partnership*, Japan has provided funds through the Japanese International Corporation Agency amounting to €8 billion to 52 partner countries including Botswana for their efforts in the environmental issues and this project was funded as part of the strategy. Furthermore, in October 2017, the Botswana Power Corporation signed a power purchase agreement with private entities to electrify 20 villages in rural areas which are mainly far from the grid using distributed network solar plants in the next 12 months. This is a positive step towards a greener environment by increasing Botswana's green energy—up to 25% by the year 2025.

Department of Energy Affairs, Ministry of Minerals, Energy and Water Affairs
The Department of Energy is responsible for the formulation, regulation, technical implementation of projects, direction and coordination of the national energy policy. The main focus of the Energy Policy is to increase the contribution of renewable energy to the country's energy needs. The policy also seeks to provide affordable, environmentally friendly and sustainable energy services in order to promote social, economic and sustainable development.

The Government is committed to exploring renewable energy, especially solar energy which provides clean energy to compliment the coal-based energy sources which are currently being used to provide electricity. For example, the use of solar energy will reduce Botswana's energy-related carbon dioxide emissions by promoting renewable and low greenhouse gases technologies. It is reported that Botswana has abundant solar energy, receiving over 3200 h of sunshine per year, with an average insolation on a flat surface of 21 MJ/m^2 per day. This rate of irradiation is one of the highest in the world. Solar energy is recognised as a promising renewable energy source in Botswana and it is currently used for water heating, refrigeration and lighting. However, its current contribution to the national energy consumption is insignificant. In order to achieve sustainable energy for all, the government has set up two organisations which are dealing with issues of renewable and clean energy.

Botswana Innovation Hub

Botswana Innovation Hub has a centre called Clean Technology, whose mandate is focussed on catalysing activities related to clean technologies, energy and environmental research and development and commercial activities within these areas. The Centre's emphasis is on sustainability and environmental protection in renewable energy, cleaner coal, water conservation and waste management.

Botswana Institute for Technology Research and Innovation

The Botswana Institute for Technology Research and Innovation, Energy Division focuses on needs based research, development and adoption of Clean Energy Technologies for Botswana, as well as optimisation of existing ones. The key areas under consideration are

- Solar powered solutions;
- Biomass technologies, and
- Optimisation of energy systems.

1.2.5 Sustainable Energy for All in Europe

European Union's policy regarding efficiency and renewable' is already running since several decades, it is nowadays regulated by the Directive 2009/29/EC of the European Parliament and of the Council, which builds on the commitment the European Council made in March 2007 to reduce the overall greenhouse gas emissions of the Community by at least 20% below 1990 levels by 2020 and by at least 50% below their 1990 levels by 2050.

The Directive 2009/29/EC prescribes

- To develop renewable energies to meet the commitment of the Community to using 20% renewable energies by 2020;
- As well as to develop other technologies contributing to the transition to a safe and sustainable low-carbon economy and to help meet the commitment of the Community to increase energy efficiency by 20% by 2020.

These goals are not directly comparable to those of the Sustainable Energy for All Global Action Agenda since they have different references, for example, regarding target years. Nevertheless, they show a common effort towards energy efficiency and renewable energies deployment.

As part of the commitment to achieving the objectives of the Sustainable Energy for All Initiative the European Commission announced on 16 April 2012 the Energizing Development Initiative,[5] which will provide developing countries with the support they need to assist them in providing access to sustainable energy. With the help of the EC, developing countries that sign on to the initiative will have the opportunity to adopt cleaner, more efficient technology from the start, leapfrogging technologies and infrastructure that developed countries established in the past.

The goal of the initiative is to provide energy services to 500 million people by 2030, by empowering developing countries through programme elements such as

- The creation of a world-class Technical Assistance Facility, drawing upon EU experts to develop technical expertise in developing countries;
- A focus on refining, expanding and improving energy-related innovative financial instruments and risk guarantee schemes in developing countries in order to unlock greater private investment;
- An effort to mobilise an additional several hundred million Euros to support concrete new investments in sustainable energy in developing countries, with the goal of leveraging even greater flows of additional investment from the private sector.

Attention should be given to the implementation of the initiative to avoid dependence phenomena regarding technologies, know-how or suppliers and to avoid the risk to exploit local renewable energies only to feed the European market.

1.3 Defining Access to Energy

A key issue in the transition towards a sustainable society is the access to modern fuels/energy for cooking [6, 7]. We know that worldwide 2.7 billion people access energy through traditional biomass, i.e. traditional three-stone wood fires for cooking. This habit carries problems around healthy and involves a gender issue. The fumes of burning fuels are a death killer for low-income people that do not have other modern fuels or energy sources for cooking. In poor context, 4000 premature deaths everyday are due to biomass fumes that is 1.5 million a year, they kill more than malaria. Furthermore, women and children make several kilometres a day to collect wood.

[5]Energizing Development Initiatives is promoted by the United Nations, aiming to provide 500 million people in the developing world with the support they need to gain access to sustainable energy. More at https://sustainabledevelopment.un.org/partnership/?p=601.

This is one of the main astonishing problems, but more in general, lack of access to energy hampers the provision of basic services such as health care, security and education [7].

Some numbers related to Energy Access [6, 7]:

- 1.2 billion people worldwide lack access to electricity;
- Furthermore, 1 billion do not have reliable access to electricity.

Who are those living without electricity?

- More than 95% are in the sub-Saharan Africa and low-income Asia countries;
- 80% of the world total are in rural areas.

Therefore, access to energy may strongly contribute to reducing inequality and poverty. Energy is an essential input to almost every good and service provided in the current economies. Energy services have a profound effect on productivity, health, education, food and water security, and communication services.

Modern fuels for cooking and heating relieve women from the time-consuming drudgery and danger of travelling long distances to gather wood. Electricity enables children to study after dark. It enables water to be pumped for crops, and foods and medicines to be refrigerated.

The World Energy Outlook 2015 highlights that access to energy also involves consumption of a specified minimum level of electricity, and the amount varies based on whether the household is in a rural or an urban area. The initial threshold level of electricity consumption for rural households is assumed to be 250 kilowatt-hours (kWh) per year and for urban households it is 500 kWh per year [7]. The higher consumption assumed in urban areas reflects specific urban consumption patterns. Both are calculated based on an assumption of five people per household. In rural areas, this level of consumption could, for example, provide for the use of a floor fan, a mobile telephone and two compact fluorescent light bulbs for about five hours per day. In urban areas, consumption might also include an efficient refrigerator, a second mobile telephone per household and another appliance, such as a small television or a computer.

Another important issue that defines access to energy is linked to the affordability of supply and legality of connection, which represent several problems, especially in low-income countries. Illegal connections are mostly in precarious housing, which increases the insecurity of families that live inside them.

Given the complexity and multiple variables which have an impact on defining energy access, SE4All's Global Tracking Framework (GTF) 2013 report introduced multi-tier frameworks for measuring it. It is divided into three areas of energy use: (i) households, (ii) productive engagements and (iii) community facilities, that together are termed as the *locales of energy access* (World Bank, 2015). Further, the model proposes indexes, organised hierarchically, that include, under the umbrella of the *overall energy access index,* the following indexes:

- Index of household access to energy, which includes energy for electricity, cooking and heating;
- Index of access to energy for productive engagements;
- Index of access to energy for community facilities, such as street lighting, health facilities, community buildings and public offices (World Bank 2015).

Considering the index of household access to energy, we can go in depth in the provision of cooking facilities. They can be used without harm to the health of those in the household and which are more environmentally sustainable and energy efficient than the average biomass cookstove currently used in low-income countries.

What we know is that the energy system we have now, mostly based on fossil fuels and centralised generation system, is whatever but not sustainable, neither in economic terms nor environmental, nor in social terms. Therefore, it is clear that we need to undergo a paradigm shift in the way we produce, supply, use and dispose of the energy.

Indeed, Distributed Renewable Energy (DRE) generation is understood by many authors [10, 15], UN [21], [1, 4, 11, 25], IRENA [9] as the paradigm shift needed in the energy sector for a Sustainable Energy for All.

The transition towards DRE is introduced in the next section.

1.4 Distributed Renewable Energy: A Key Leverage Towards Sustainable Energy for All (SE4A)

The **Distributed Renewable Energy (DRE) generation** could be defined as '*a small-scale generation units harnessing renewable energy resources (such as sun, wind, water, biomass and geothermal energy), at or near the point of use, where the users are the producers—whether individuals, small businesses and/or a local community. If the small-scale generation plants are also connected with each other (to share the energy surplus), they become a Renewable Local Energy Network, which may in turn be connected with nearby similar networks*'.

The main environmental benefits of a DRE are, since they use non-exhaustible resources, they have low greenhouse gas emissions, they produce low environmental impact for extraction, transformation and distribution (low-energy transmission losses) compared to non-renewable centralised energy generation units.

The main socioethical and economic benefits are due to the small scale of generation units that require small economic investment, are easy to instal, maintain, manage and allow individuals and local communities to instal/manage them, thus leading to democratisation of access to resources, which improves quality of life and enhances local employment and dissemination of competences.

One of the most committed and known researchers on the sustainable energy topic is Jeremy Rifkin, who is speaking about the *third industrial revolution* [15] and his core idea claims 'the creation of a renewable energy regime, loaded by buildings, partially stored in the form of hydrogen, distributed via an energy internet—a smart intergrid—and connected to plug in zero emission transport'.

In accordance with his thinking, it is possible to set up some useful features or needed pillars for the third industrial revolution [15]:

- Shift to renewable energy (solar, wind, hydro, geothermal, ocean waves and biomass);
- Transform buildings as power plants;
- Deploy hydrogen and other storage technologies in every building and throughout the infrastructure to store intermittent energies;
- Use internet technology to transform the power grid of every continent into an energy-sharing inter-grid that acts just like the internet;
- Transition the transport fleet to electric, plug-in and fuel cell vehicles that can buy and sell electricity on a smart continental interactive power grid.

These pillars, using Distributed Renewable Energy (DRE) systems, represent promising steps towards Sustainable Energy for All.

Both the DRE and their sustainability benefits are fully described in Chap. 2.

1.5 Sustainable Product-Service Systems Applied to Distributed Renewable Energy: An Introduction

Since the end of 90s, Sustainable Product-Service System (S.PSS) has been studied as a promising opportunity for sustainability [2, 3, 5, 13, 17–19, 23, 24, 27–28]. S. PSS are defined [26] as 'offer models providing an integrated mix of products and services that are together able to fulfil a particular customer demand (to deliver a 'unit of satisfaction'), based on innovative interactions between the stakeholders of the value production system (satisfaction system), where the economic and competitive interest of the providers continuously seeks environmentally and socioethically beneficial new solution'.

Sustainable Product-Service System has been considered within the LeNSes project as a promising model for the diffusion of Distributed Renewable Energies in low- and middle-income contexts.

In fact, the following is the outcome of the multiregional research carried out during the LeNSes project: '*A S.PSS applied to DRE is a promising approach to diffuse sustainable energy in low/middle-income contexts (for All), because it reduces/cuts both the initial (capital) cost of DRE system purchasing (that may be unaffordable) and the running costs for maintenance, repair, upgrade,* etc. *(that may cause interruption of use), while increasing local employment, related skills and entrepreneurship, as well as fostering for economic interest the design of low environmentally impacting DRE products, resulting in a key leverage for a sustainable development process aiming at democratizing the access to resources, goods and services*'.

The articulation and characteristics of such promising outcome are presented in this book, together with the role the designer should play to develop them, i.e. the new discipline of System Design for Sustainable Energy for All (SD4SEA), namely the design of S.PSS applied to DRE.

The following section examines one case study to further introduce S.PSS applied to DRE.

Solarkiosk, Africa

The Solarkiosk AG (German company) targets local intermediaries to manage and guarantee the provision of energy services in rural areas of Kenya, Rwanda and Tanzania. Solarkiosk designs, instals and owns the E-Hubb, a charging station provided with solar panels and various equipment and products depending on the location, such as computer, printer, solar lanterns and fans. A local intermediary is responsible (with a maximum of 5 collaborators) of the local E-Hubb, where (s)he provides a wide range of energy services such as Internet connectivity, copying, printing and scanning, etc. *Customers pay per use, e.g. pay to print, or they can buy some offered products or food products. Local intermediary receives training for management, selling and accountability of the E-Hubb, as well as to solve basic maintenance and repair. Currently, as new market segment for the E-Hubb, Solarkiosk is offering energy connection to local shops, thus entailing more favourable conditions for the local economy, e.g. in food shop, the access to reliable energy can power refrigerators to keep goods.*

For the customer, the opportunity to obtain her/his result is given from the small payment (s)he can give for each use. For example, to send out an email, the customer pays a fixed amount, without making any initial investment to buy a computer to send it, neither paying unexpected costs in case an upgrade or repair of the computer if needed. In the case of products which are sold to the customer, e.g. solar lantern, the local intermediary is in charge to solve technological problems related to the product as additional service, without extra costs for the customer. The products, both used in the E-Hubb and sold, are certified, so that quality and efficiency are ensured, both for the local intermediary to work on them and for the clients. For the local intermediary, the training courses can increase their competencies and future opportunities for job career. On the side of the Solarkiosk AG Company, they had the opportunity to enter the untapped market of rural areas. In fact, even though all customers have limited power purchase, the possibility to cover high numbers still gives them margins of return. Finally, on the environment, the use of Renewable Energy solutions, both the E-Hubb as well as the products such as lanterns and efficient cookstoves, can increase the quality of the given results, while reducing their environmental impact.

The Distributed Renewable Energy (DRE) systems and the Sustainable Product-Service System (S.PSS) win-win models are, respectively, introduced in Chaps. 2 and 3 of this book. The S.PSS applied to DRE approach is presented in Chaps. 4 and 5. Consequently, the new key role for designers defined as System Design for Sustainable Energy for All (SD4SEA) is presented in Chap. 6, as the way for the designers to contribute to the transition towards a sustainable society.

Chapters 7 and 8 are dedicated to the method and the tools to support the designers in their practice (developed by the LeNSes project partners), together with on-field experiences conducted during the same project.

References

1. Barbero S, Pereno A (2013) Systemic energy grids: a qualitative approach to smart grids. J Record 6(4):220–226
2. Brandstotter M, Haberl M, Knoth R, Kopacek B, Kopacek P (2003) 'IT on demand—towards an environmental conscious service system for Vienna (AT)', Third International Symposium on Environmentally Conscious Design and Inverse Manufacturing—EcoDe- sign'03 (IEEE Cat. No.03EX895)
3. Baines TS, Lightfoot HW, Evans S, Neely A, Greenough R, Peppard J, Roy R, Shehab E, Braganza, Tiwari A, Alcock JR, Angus JP, Bastl M, Cousens A, Irving P, Johnson M, Kingston J, Lockett H, Martinez V, Michele P, Tranfield D, Walton IM, Wilson H (2007) State-of-the-art in product-service systems (Cranfield, UK: Innovative Manufacturing Research Centre, Cranfield University)
4. Colombo E, Bologna S, Masera D (2013) Renewable energy for unleashing sustainable development. Springer, United Kingdom
5. Goedkoop M, van Halen C, te Riele H, Rommes P (1999) Product Service Systems, Ecological and Economic Basics, report 1999/36 (the Hague: VROM)
6. International Energy Agency (IEA) (2011) World Energy Outlook. IEA Publications, France
7. International Energy Agency (IEA) (2015) World Energy Outlook. IEA Publications, France
8. International Energy Agency (IEA) (2016) World Energy Outlook. IEA Publications, France
9. International Renewable Energy Agency (IRENA) (2016) Renewable energy statistics. International Renewable Energy Agency, Abu Dhabi
10. Johansson TB et al (2002) Energy for sustainable development—a policy agenda. IIEEE, Lund
11. Klein N (2014) This changes everything. Capitalism vs the Climate. Simon & Schuster, New York
12. Meadows DH, Meadows DL, Randers J (2004) Limits to growth: the 30-year update. Chelsea Green Publishing
13. Mont O (2002) Functional thinking: The role of functional sales and product service systems for a functional based society, research report for the Swedish EPA (Lund, Sweden: IIIEE Lund University)
14. Nilsson M (2012) Sustainable energy for all: from basic access to a shared development agenda, Carbon Manag 3(1):1-3. Future Science Ltd
15. Rifkin J (2011) The third industrial revolution. How lateral power is transforming energy, the economy, and the world. P. Macmillan, New York
16. Rogelj J McCollum DL Riahi K (2013) The UN's 'sustainable energy for All' initiative is compatible with a warming limit of 2 °C, Nat Clim Change 3:545–551
17. Stahel W (2001) 'Sustainability and Services'. In: Charter M, Tischner U (eds) Sustainable Solutions—Developing products and services for the future (Sheffield, UK: Greenleaf Publishing)
18. Tischner U, Ryan C, Vezzoli C (2009) Product- Service Systems. In: Crul M, Diehl JC (eds) Design for Sustainability a global guide. Modules. United Nations Environment Program (UNEP)
19. UNEP (2002) Product-Service Systems and Sustainability: Opportunities for Sustainable Solutions (Paris: United Nations Environment Programme, Division of Technology Industry and Economics, Production and Consumption Branch)

20. United Nations World Commission on Environment and Development (WCED) (1987) Our common future. Oxford University Press, UK
21. United Nations (2011) Sustainable energy for all: a vision statement (by Ban Ki-moon secretary-general of the United Nations). United Nations
22. United Nations (2015) Transforming our world: the 2030 agenda for sustainable development. United Nations
23. van Halen C, Vezzoli C, Wimmer R (eds) (2005) Methodology for Product Service System. How to develop clean, clever and competitive strategies in companies (Assen, Netherlands: Van Gorcum)
24. Vezzoli C (2011) 'System Design for Sustainability: The new research frontiers'. In: Haom-ing Z, Korvenmaa P, Xin L (eds) Tao of Sustainability: Strategies in a globalisation context, Proceedings, Academy of Arts and Design, Tsinghua University, Beijing, 27-29 October, 2011
25. Vezzoli C, Delfino E, Amollo Ambole L (2014) System design for sustainable energy for all. A new challenging role for design to foster sustainable development [online]. Available at http://dx.doi.org/10-7577/formakademisk.791
26. Vezzoli C, Ceschin F, Diehl JC (2015) 'The goal of sustainable energy for all. SV J Cleaner Prod 97:134–136
27. White AL, Stoughton M, Feng L (1999) Servicizing: The Quiet Transition to Extended Product Responsibility (Boston, USA: Tellus Institute)
28. Zaring O, Bartolomeo M, Eder P, Hopkinson P, Groenewegen P, James P, de Jong P, Nijhuis L, Scholl G, Slob A, Örninge M (2001) Creating eco-efficient producer services. Gothenburg Research Institute, Gothenburg

Chapter 2
Distributed/Decentralised Renewable Energy Systems

2.1 Distributed/Decentralised Renewable Energy: Sustainability

In the previous chapter, we introduced that Distributed Renewable Energy (DRE) is the most promising model to bring sustainable energy to All. Figure 2.1 schematizes the paradigm shift from non-renewable/centralised energy generation systems to renewable/distributed energy generation unit. Let us see better why DRE is environmentally, socioethically and economically sustainable compared with the dominant centralised and non-renewable energy generation systems.

Environmental benefits of DRE
If we look at centralised and non-renewable systems, namely, large-scale plants using fossil fuels as oil and coke, they are environmentally unsustainable because they are based on exhausting resources, so forth fastening resources depletion. Furthermore, these exhausting resources result in high greenhouse gases emission (CO_2 emissions), through several processes along their life cycle, which determine global warming. Finally, they are responsible for other pollution problem during extraction and transportation processes due to their linking.

If we now look at renewable and distributed resources, such as small-scale solar and wind generation units, they are more environmentally sustainable because they use locally available and renewable energy sources, thus resulting in a reduced environmental impact compared to the various processes of extraction, transformation and distribution of fossil fuels. Furthermore, they have much lower greenhouse gases emissions in use. To conclude, compared to centralised systems, local energy production and distribution increase reliability and reduce distribution losses.

Socioethical and economic benefits of DRE
Centralised systems are unsustainable even in socioethical and economic terms. This comes because, due to the composition of oil and coke, they are very complex

C. Vezzoli et al., *Designing Sustainable Energy for All*,
Green Energy and Technology, https://doi.org/10.1007/978-3-319-70223-0_2

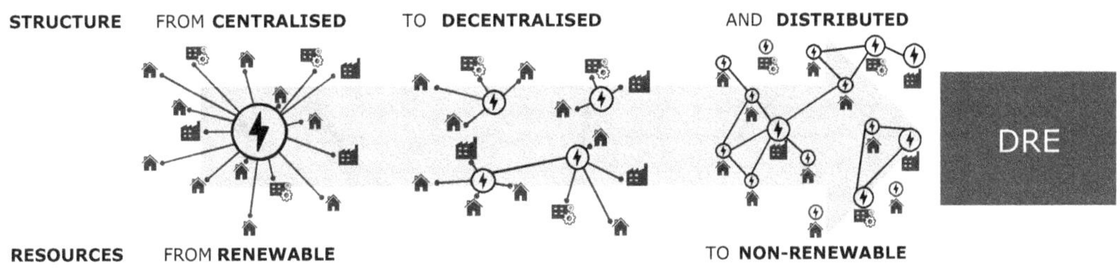

Fig. 2.1 Paradigm shift from non-renewable/centralised energy generation systems to renewable/distributed ones. *Source* designed by the Authors

to be extracted, refined and distributed. Indeed, these processes require very expensive and large-scale centralised structures, which limit the possibilities of direct and democratised access to energy production and consumption. In history, individuals had low power over their own destiny which led to a widened gap (in terms of inequality) between rich and poor [10], which has been pursued in time perpetuating a centralised energy production.

In contrast, the main advantage of DRE systems is related to their reliability and resilience. In fact, because of their distributed architecture, DRE systems can easily cope with individual failures, since each energy-using node can be served by multiple energy production units (while a fault in a centralised system might affect the energy distribution in the whole system). For example, small generation units for energy production are manageable by small economic entities, where the user can become prosumer (producer + consumer) and the generation units could be connected in a micro energy network, potentially connected with a global network. On this perspective, DRE systems could enable a democratisation of energy access, thus fostering inequality reduction, community self-sufficiency and self-governance. It has been estimated that Distributed Renewable Energies (DRE) has the potential to enable energy access to more than 1 billion by 2025 [12].

2.2 Distributed/Decentralised Renewable Energy Systems: Structures and Types

In the transition from centralised to decentralised and distributed energy systems, there are two well-characterised elements:

- **System Structure**: regarding the configuration of the actors involved in the energy system;
- **Type of Energy Sources**: regarding the nature of the resources, covering from non-renewable to renewable energy sources.

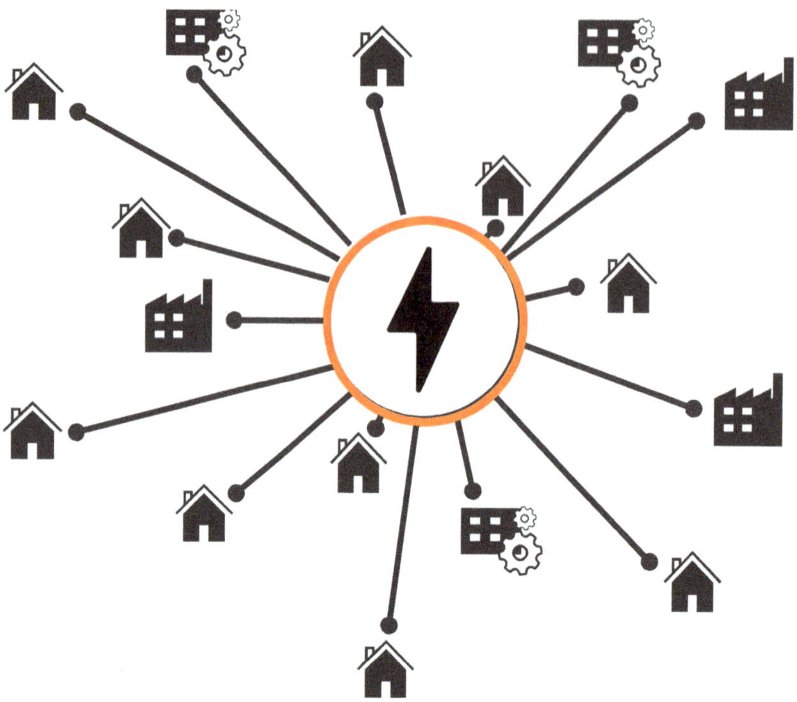

Fig. 2.2 Centralised energy system. *Source* designed by the Authors

Concerning the **System Structure,** we can distinguish the following three main types.[1]

Centralised energy systems could be defined as *large-scale energy generation units (structures) that deliver energy via a vast distribution network, (often) far from the point of use* (Fig. 2.2).

Decentralised energy systems could be defined as characterised by small-scale energy generation units (structures) that deliver energy to local customers. These production units could be stand-alone or could be connected to nearby others through a network to share resources, i.e. to share the energy surplus. In the latter case, they become locally decentralised energy networks, which may, in turn, be connected with nearby similar networks (Fig. 2.3).[2]

Distributed energy system could be defined as *small-scale energy generation units (structure), at or near the point of use, where the users are the producers—whether individuals, small businesses and/or local communities. These production units could be stand-alone or could be connected to nearby others through a network to share, i.e. to share the energy surplus. In the latter case, they become locally distributed energy networks, which may, in turn, be connected with nearby similar networks* (Fig. 2.4).

---

[1]The definitions given here are the ones adopted by the LeNSes project.

[2]In some classifications (e.g. Colombo et al. [2]) decentralised systems, differently than in the LeNSes approach, are only individual and isolated systems.

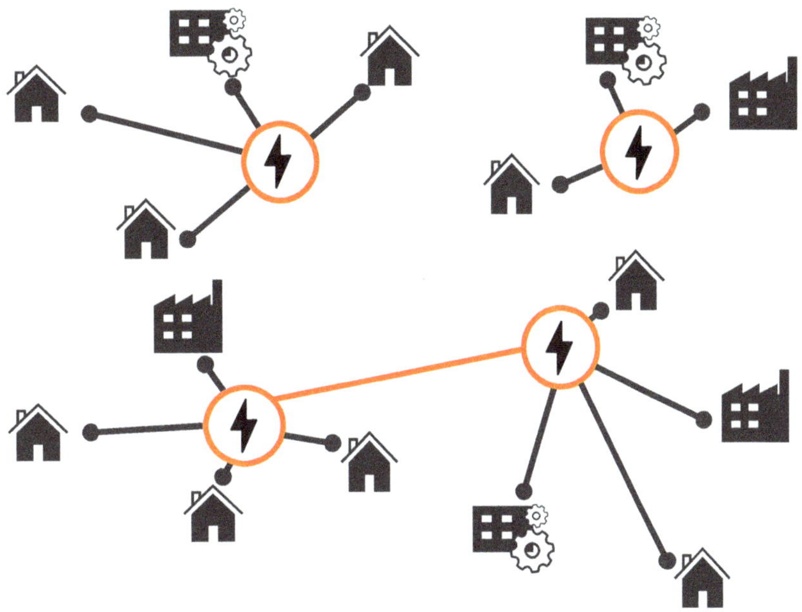

Fig. 2.3 Decentralised energy system. *Source* designed by the Authors

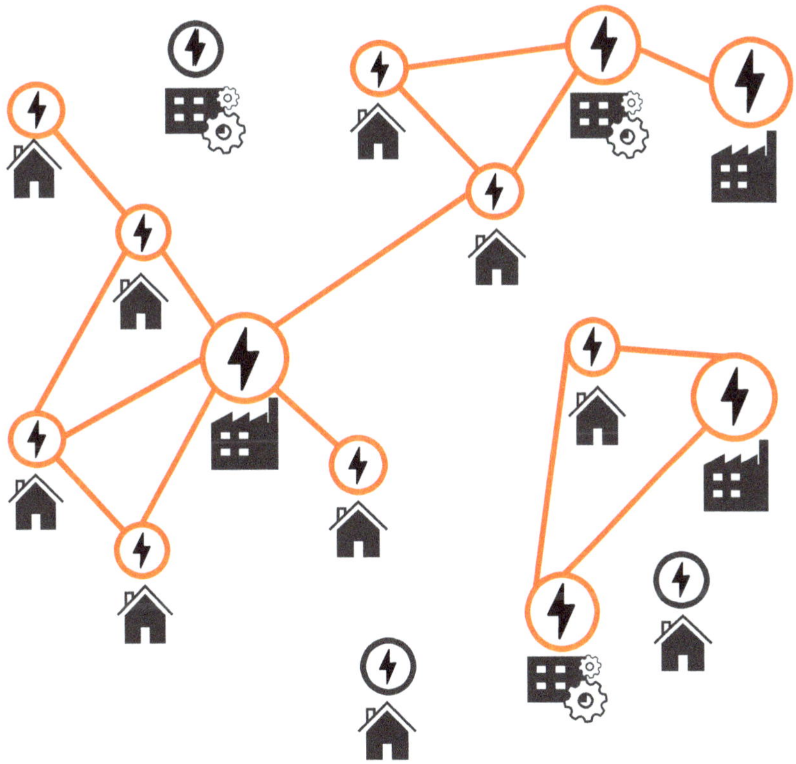

Fig. 2.4 Distributed energy system. *Source* designed by the Authors

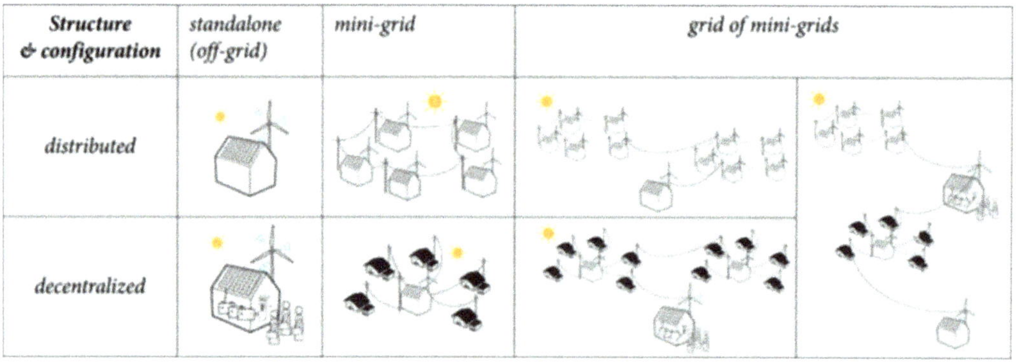

Fig. 2.5 Distributed/decentralised energy. System structure and configurations. *Source* designed by the Authors

Given the above structures, the below diagram presents various types of possible configurations (Fig. 2.5).

2.3 Renewable Energy Systems Types

An explanation is needed on the renewability of resources. On one side, we can recognise the nature of the resource, considering the kind of transformation needed to make them usable. Some exhaustible resources, such as oil, are available as fossil hydrocarbons, but we can only use them after extraction and converting them into heat, electricity and so on. These extraction and conversion processes imply having, as it was highlighted before, large-scale centralised plants. With renewable resources, this transformation processes could be relatively simpler. The simplest example comes out with the sun: it is freely available and it can directly be used in the form of heat for cooking and even for house heating.

On the other side, we can characterise resources based on their capability of regeneration against the anthropic consumption rate. It means that this resource could be continuously available for its use, under the condition that it is correctly managed. Wood represents a typical case whereas renewability depends on this. The same type of wood could be renewable or not depending on how its growth is being planned and controlled. Once again, we cannot define a renewable resource without mentioning the context in which it is produced and consumed. What can be 'renewable' on one side of the world, with given natural sources, culture even political situation, could be considered 'non-renewable' in other locations. Because of that, recognising the context is one the pillars towards creating a distributed renewable energy system. The renewable energy sources are the following: sun, wind, water, biomass and geothermal energy. An explanation of the main resources is provided in the next paragraphs.

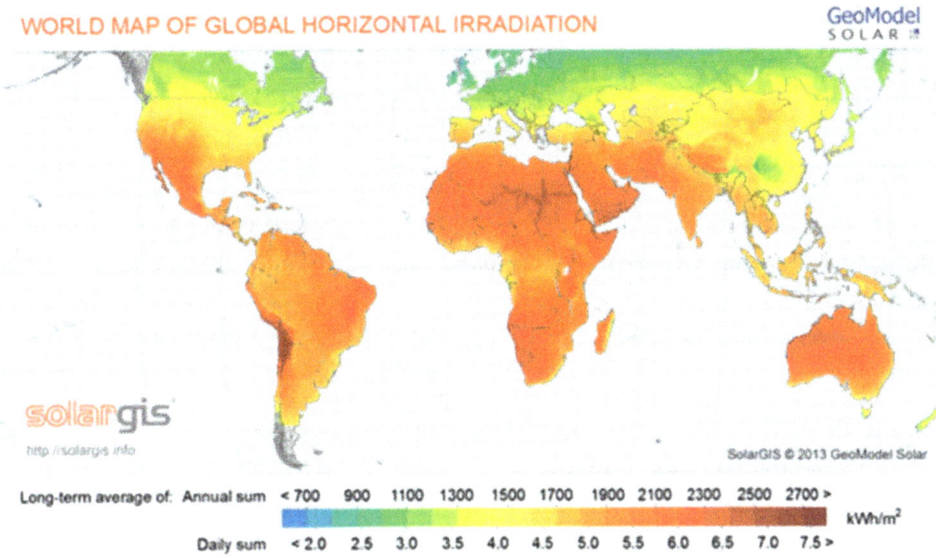

Fig. 2.6 World map solar horizontal irradiation. *Source* https://solargis.com/legal/terms-of-use-for-ghi-free-maps/

2.3.1 Solar Energy

Solar Energy is the most abundant of renewable energies, and it is available at any location, with higher values/yields closer to the Equator, e.g. 1400–2300 kWh/m^2 in Europe and US and around 2500 kWh/m^2 in Tanzania, East Africa [11]. The total solar irradiation of the sun is about 50 million Gigawatt (GW) (Fig. 2.6).

The value of radiation is influenced by seasonal climatic variations: it is higher during warmer months than in cold months and usually is higher during the dry season than rainy season.

Nowadays several studies and databases are available to obtain a first estimation of the annual PV plant energy production for a selected location. Two examples of free database are as follows: Photovoltaic Geographical Information System (PVGIS)[3] provides a map of solar energy resource and assessment of the electricity generation from photovoltaic systems in Europe, Africa and South–West Asia. It provides information related to distributed generation or stand-alone generation in remote areas; IRENA's Global Atlas[4] provides maps of resources and support tools to evaluate the technical potential of both solar and wind energy. It includes socio-economic data. When no data are available, field measurements of solar radiation can be made using solar radiometers even though affection from external factor can be expected.

[3]Photovoltaic Geographical Information System, http://re.jrc.ec.europa.eu/pvgis/.

[4]Global Atlas for solar and wind, www.irena.org/globalatls/.

Solar Technologies

There are two main solar energy technologies: solar photovoltaic systems which use solar irradiation to produce electricity, and solar thermal systems that make use of the sun's heat, e.g. in solar cooking and solar water heating.

Solar Photovoltaic Systems (SPS) convert the energy from the sun using solar cells: the PV effect related to the electromotive force is generated under the action of light in the contact zone between two layers of semiconductor material usually silicon-based.

Solar Photovoltaic Systems (SPS) typically are composed of the following components:

- Photovoltaic Cell/Module/Array: to convert solar energy into electric energy through the photovoltaic effect;
- Charge Controllers: to protect and regulate the charging of batteries, the charge controller interrupts the photovoltaic current when the battery is charged;
- Rechargeable Battery bank: to store the surplus of solar energy if not connected to the grid. Types of batteries are: deep cycle lead acid, gel, lithium polymer, lithium ion and NiCad (Nickel Cadmium), and these have a range between 12 and 48V, where the higher the voltage the better the efficiency;
- Inverter: to convert the DC from the photovoltaic modules in AC (necessary for products such as domestic appliances, computers, cars and urban lights). There are two different types: converts DC to AC; runs at 120VAC or 240VAC appliances;
- Breaker box: to distribute electrical current to the various circuits (if grid connected);
- Electric metre: to measure electric energy delivered to their customers (if connected to a network) for billing purposes;
- Wires/cables.

If the dimension of the SPV is limited (less than 100 W), the inverter can be avoided, thus avoiding conversion losses. On the other side, to reach a higher output capacity, a certain number of modules are combined to form a field or array. This example shows the solar high degree of flexibility and scalability of Solar Photovoltaic Systems (SPV), able to power from small lanterns up to mini-grid systems connecting more energy generator units (some hundreds kWp). When considering microgrid systems, about 50–60% of the total cost is due to the solar PV array, while battery bank accounts for about 10–15% and power conditioning unit for 25–35%.

Solar thermal technology converts solar radiation into renewable energy for heating and cooling using a solar thermal collector. Heat from the sun's rays is collected and used to heat a fluid that will drive the production of energy for heating/cooling. Produced heat can be used to heat water for hygiene and health, or for space heating/cooling (e.g. solar driers and greenhouses).

Solar thermal heating systems are typically composed of the following components: solar thermal collectors, a storage tank and a circulation loop.

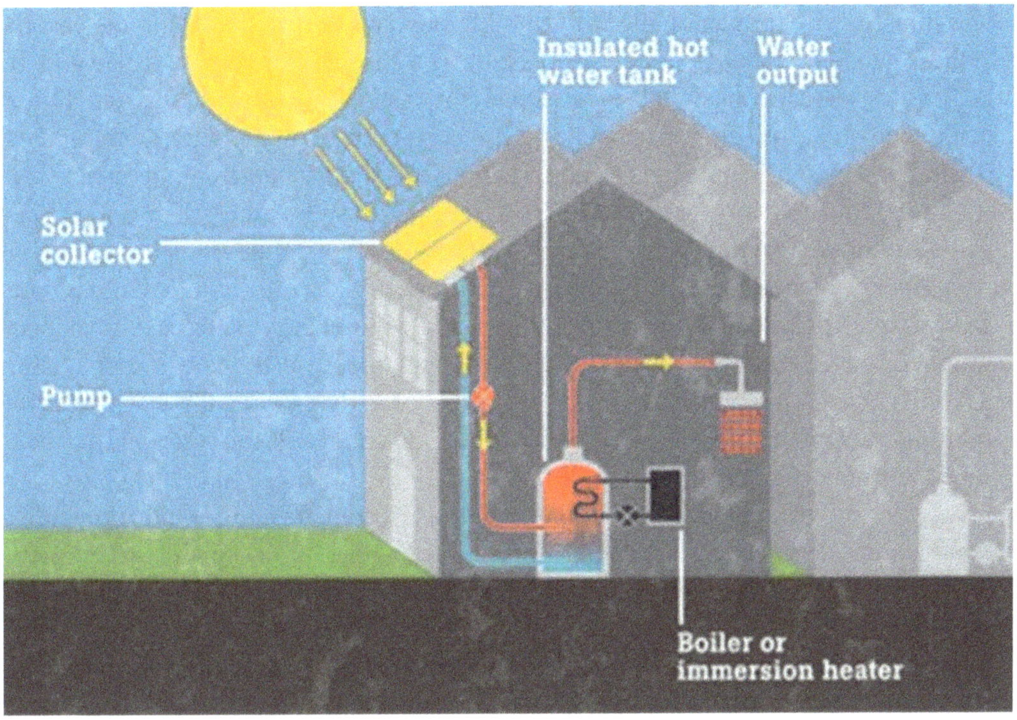

Fig. 2.7 Solar heaters components. *Source* www.ashden.org

The solar thermal collector is composed of:

- An absorber metal, such as copper/steel covered with chromo, alumina–nickel and Tinox. These materials give high conductivity, high absorptivity and low emissivity;
- An insulating system that provides a low thermal conductivity to make the whole system resistant to high temperature. It can be made from rock wool, polyurethane foam, polystyrene and others;
- Circulating tubes are constructed from metals with good conductivity;
- Transparent coverage reduces heat losses and maximises the efficiency of the collector (Fig. 2.7).

2.3.2 *Wind Energy*

Wind power is extremely site-specific. The energy produced by a wind turbine along the year depends on the average wind speed at the installation site (to achieve economic sustainability, it is required an average wind speed of 4–5 m/s along the year) and is highly influenced by geography and barriers that might obstacle for the passage of wind through the turbines.

Obviously, wind power changes during the day and during the different seasons. For these reasons, data on local wind resources throughout the year need to be collected to select most suitable locations for wind turbines installation. Direct measurements can be taken by installing meteorological towers with anemometers and wind vanes to measure speed and directions. Secondary data can be taken from other measuring meteorological or airport installations, together with appropriate calculation models. A further possibility is provided by online databases, such as the previously mentioned IRENA's Global Atlas for solar and wind. Online databases can offer only very limited information for wind energy, since, as it has been mentioned, the average wind speed is highly dependent on the specific characteristics of a chosen area. Furthermore, as wind resource maps typically evaluate wind conditions at 50 m height, the information obtained can result too different for those relevant for small wind turbines.

The working principle of wind energy consists of transforming wind force into a mechanical or electrical one. A Wind Power Generator (WPG) converts the kinetic energy of the wind, through rotor blades connected to a generator, into electric power. In the case of an air-generator, the force of the wind turns the blades, converting the energy of the wind into mechanical energy of the rotating shaft. This shaft is then used to turn a generator to produce electricity or to operate a mechanical pump or grinding mill.

The main wind power system components are as follows:

- A rotor, or blades, which convert the wind's energy into rotational shaft energy;
- A nacelle (enclosure) containing a drive train, usually including a gearbox and a generator;
- A tower, to support the rotor and drive train;
- Electronic equipment such as controls, electrical cables, ground support equipment and interconnection equipment.

With similar components, there are two basic designs of wind electric turbines:

- Horizontal-axis (propeller-style) machines;
- Vertical axis, or 'egg-beater' style.

Horizontal-axis wind turbines are most common today.

The price depends on the size, material and construction process. Costs of Small Wind systems include turbine and components: tower or pale, battery storage, power conditioning unit, wiring and installation, as well as maintenance: turbine requires cleaning and lubrication, while batteries, guy wires, nuts and bolts, etc. require periodic inspection. Costs depend on the cost of local spares and service.

2.3.3 Hydro Energy

Energy from water can be produced through different sources: water flow, waves or from the tide, all cases it is transformed into mechanical power or could be converted into electricity. There are three different technologies using water:

hydropower, energy from waves, energy from the tide. Currently, hydropower is a mature technology; last two are at the level of experimentations. So forth, here only hydropower will be presented.

Hydropower resources are extremely site-specific: the right combination of flow and fall is required to meet a certain electric load. Best geographical areas to instal a hydropower system are generally in presence of perennial rivers, hills or mountains, but since a river flow can vary greatly during the seasons, a single measurement of instantaneous flow in a watercourse is not enough, it is important to gather detailed information to estimate energy production potential. Moreover, also the evaluation of the best site is required. For some areas, general data about water resources assessment can be found on Info hydro, a database provided by the World Meteorological Organization. However, in most cases, data for the site of interest are not available, or a more accurate estimation is strictly necessary. For these reasons, a direct evaluation is required.

To measure the flow, there exist several methods. A brief description of the two most common methods is given here below.

- Velocity-area method: this method is suitable for medium-sized rivers. The evaluation of the stream is obtained by measuring the cross-sectional area of the river and the speed of the water;
- Weir method: for small rivers, a temporary weir can be built. This is a low obstacle across the stream to be gauged with a notch through which all the water may be channelled. Water flow measurement is obtained by a measurement of the difference in level between the upstream water surface and the bottom of the notch;

Hydropower plants transform kinetic energy into mechanical energy with a hydraulic turbine. The power available in a river or stream depends on the rate at which the water is flowing, and the height (head) that falls. Mechanic energy drives devices or is converted into electric energy via an electric generator. Electricity production is continuous, as long as the water is flowing.

The most typical hydropower system is composed of the following elements:

- Weir and intake channel: where water is diverted from the natural stream, river or perhaps a waterfall;
- Forebay tank: artificial pool to contain water;
- Penstock: canal to bring water to the turbine;
- Power group: the turbine converts the flow and pressure of the water into mechanical energy. The turbine turns a generator connected to electrical load, directly connected to the power system of a single house or to a community distribution system.

Hydropower plant costs depend on site characteristics: terrain and accessibility, (for micro-systems) the distance between the powerhouse and the loads can have a significant influence on overall capital costs; the use of local materials, local labour and pumps; operational costs are low due to high plant reliability, proven technology.

2.3.4 Biomass Energy

Bioenergy is made available from biomass, e.g. crops, residues and other biological materials that could be used to produce chemical energy, i.e. gas that could be converted into electricity. Also, transportation fuels can be produced from biomass, thus reducing the demand for petroleum products. Main transportation fuels are ethanol from corn and sugarcane, and biodiesel from soy, rapeseed and palm oil.

Biogas, a mixture of methane and carbon dioxide, is produced by breaking down biomass, particularly wet organic matter like animal dung, leftover food or human waste. The main biogas digester system is composed of the following elements:

- A large container to hold the mixture of decomposing organic matter and water (which is called slurry);
- Another container to collect the biogas;
- Opening to add the organic matter (the feedstock);
- Opening to take the gas to where it will be used;
- Opening to remove the residue.

In fixed dome biogas plants (the most common type), the slurry container and gas container are combined.

The gasification process to produce chemical energy entails a partial combustion of biomass due to the limited presence of air in the reactor. The gasification of biomass takes place in four stages:

- Drying: water vapour is driven off the biomass;
- Pyrolysis: as the temperature increases, the dry biomass decomposes into vapours, gases, carbon (char) and tars;
- Reduction: water vapour reacts with carbon, producing hydrogen, carbon monoxide and methane. Carbon dioxide reacts with carbon to produce more carbon monoxide;
- Combustion: some of the char and tars burn with oxygen from air to give heat and carbon dioxide. This heat enables the other stages of the gasification process to take place;

Figure 2.8 shows the process of gasification:

- Updraft gasifier, where biomass is loaded at the top of the gasifier and air is blown in at the bottom. This type of gasifier produces gas that is contaminated by tar and is therefore too dirty to be used in an internal combustion engine;
- Downdraft gasifier, where air is drawn downwards through the biomass. The main reactions occur in a constriction or 'throat', where the tars and volatile gases break down into carbon monoxide and hydrogen at a much higher temperature than in an updraft gasifier. The throat is usually made from ceramic to withstand this temperature. Downdraft gasifiers produce cleaner gas.

The cost of biogas plants varies greatly from country to country, depends on the costs of both materials (brick, concrete and plastic) and labour that can be very

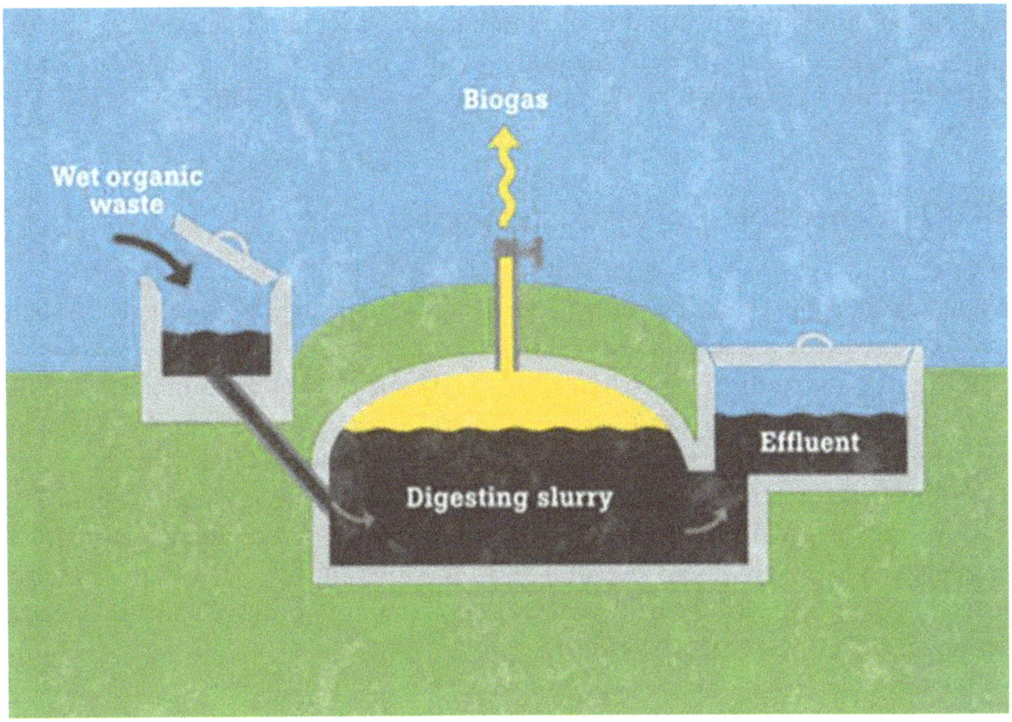

Fig. 2.8 Process of gasification. *Source* www.ashden.org

different by context. The cost per cubic metre of digester volume decreases as volume rises. Using plastic or steel to prefabricate biogas plants usually increases the material cost but can substantially reduce the labour needed for installation as well as the lifetime (compared to flexible bags). Biomass gasification is not suitable for home-based solutions due to the low efficiency and high quantity of biomass needed compared to the chemical energy produced.

2.3.5 Geothermal Energy

Geothermal energy can be found in rocks in fluids that circulates underground. The main use of this kind of renewable energy is the direct use of its heat, e.g. to heat buildings, to grow plants in greenhouses, to dry crops, to heat water at fish farms and several industrial processes, or the conversion of such heat into electricity for different purposes.

Geothermal energy requires a heat pump, an air delivery system (ductwork) and a heat exchanger—a system of pipes buried in the shallow ground near the building. The heat pump converts the low temperature of geothermal energy into thermal energy with a higher temperature, thus exploiting the physical property of fluids to absorb and to release heat when they vaporise or condense, respectively. Main technologies using geothermal energy are the geothermal heat pumps, which use the shallow ground to heat and cool buildings; the geothermal electricity production, which generates electricity from the earth's heat; and the geothermal direct use, which produces heat directly from hot water within the earth.

2.4 Is Renewable Energy Zero Impact?

When talking about renewable energy and its environmental impact, there are some common myth conceptions that have to be debunked.

First, it is sometimes believed that renewable energy has zero impact. Even if renewable energy systems do not produce harmful emissions in the use stage,[5] it must be said that these systems do have an environmental impact. This is mainly related to the extraction of resources and the manufacturing processes required to produce the physical elements of the energy systems. In addition, distribution, maintenance and disposal also contribute to the total impacts. The overall impact depends on the type of energy source, the geographic location and the specific characteristics of the energy systems.

On the other hand, another myth conception, in particular in relation to PV energy systems, is that manufacturing a solar panel consumes more energy that it will ever deliver in its lifespan [6]. This is of course false. If we look at the *energy yield ratio* (the ratio of energy produced by a system during its lifespan to the energy needed to make it), PV systems generally range from 4 (for a grid-connected system in central Northern Europe) to more than 7 in Australia (ibid.).

The energy yield ratio is an interesting indicator to show the efficiency of an energy source in terms of energy returned (by the system) on energy invested (to manufacture and operate the system). Typical energy yield ratios[6] for electric power generated using common energy sources are as follows [5]. Hydroelectric power has the highest value, 84. This is followed by wind power, which has a ratio of 20. Geothermal and solar have a similar mean value, around 10. Regarding fossil fuels, coal has a ratio of around 12, while natural gas has a mean value around 7.

Although interesting, the energy yield ratio represents only one element of the picture. What this ratio does not tell us is the overall impact of using a particular energy source. For example, geothermal, solar and coal have a similar energy yield ratio, but this does not mean they have similar environmental impacts. To this end, we need to look at the impact generated considering the whole energy production chain, from exploration and extraction to processing, storage, transport, transformation and final use. For example, considering only greenhouse gases emissions, the World Energy Council [13] shows that photovoltaic, hydro and wind energy have CO_2eq emissions between around 10 and 100 tonnes per GWh of electricity.

---

[5]Even if we should also consider the impact related to maintaining the energy system (e.g. cleaning, replacing batteries or other components).

[6]Energy yield ratios change historically. Also, each individual energy system has its own specific ratio.

This is considerably lower than the emissions related to natural gas (around 400 CO_2 eq./GWh), oil (between around 650 and 800 CO_2 eq./GWh), and coal (between around 800 and 1000 CO_2 eq./GWh).

Even if renewable energy has a lower impact than fossil fuels, it is important to understand specific impacts associated with the technology used:

- Wind turbines are linked to impact on wildlife, and in particular bird and bat deaths from collisions with wind turbines, caused by changes in air pressure by the spinning turbines, as well as from habitat disruption [9]. However, as concluded in the NWCC report, these impacts do not pose a threat to species populations;
- In relation to PV cells, we need to consider the hazardous materials needed to clean the semiconductor surface. These can include, depending on the type and size of cell, hydrochloric acid, sulphuric acid, nitric acid, hydrogen fluoride and acetone [8]. Thin-film PV cells use some toxic materials not used in traditional silicon photovoltaic cells, including gallium arsenide, copper–indium–gallium–diselenide, and cadmium-telluride (ibid.). Thus, it is important to prevent exposure to workers and ensure proper disposal. Other associated impacts include land use, especially in relation to relatively big plants;
- Hydropower is associated with alteration of ecosystems, as the construction of dams is likely to influence the flow of rivers (with potentially related drained rivers and floods). This can have an impact on wildlife as well as people's activities.

2.5 Barriers to Distributed/Decentralised Renewable Energy Systems

Even though a wide range of socio-economic and environmental arguments are in favour of Distributed Renewable Energy systems (DRE), in practice there are also a series of barriers to overcome. In this perspective, a barrier to a DRE may be defined as a factor that negatively affects its adoption and subsequent utilisation which hampers its widespread diffusion [14]. Large-scale diffusion and utilization of relatively newer technologies such as DREs face barriers. These barriers may put DREs at a technical, economic, regulatory or institutional disadvantage in comparison to conventional energy systems [1]. Several scholars have identified and clustered barriers for specific renewable energy system (i.e. photovoltaic) as well as in more general for a range of DREs.

For example, Karakaya and Sriwannawit [4] conclude that the adoption of PV systems—either as a substitute for other electricity power generation systems in urban areas or for rural electrification—is still a challenging process. Although photovoltaic (PV) systems have become much more competitive, the diffusion of PV systems still remains low in comparison to conventional energy sources. They still face several barriers encompassing four dimensions: *sociotechnical, management, economic and policy*. From the economic point of view, the cost of PV systems is still generally perceived as high. In regard to the sociotechnical dimension, several studies imply that the complexity of interaction between people and PV systems can hinder the adoption. In addition, there are still several barriers related to the policy dimension and technology management. Ineffective policy measures and inappropriate management can hamper the diffusion process in a variety of contexts.

Some authors [7] identified three main barriers to the deployment of renewables in developing countries: there are, respectively, *policy and legal barriers, technical barriers and finally, financial barriers*. According to their work, the introduction and success of any renewable technology are, to a large extent, dependent on *the existing government policies*. Government policies are an important factor in terms of their ability to create an enabling environment for DREs dissemination and mobilising resources, as well as encouraging private sector investment. Specifically, the success of DREs in the Western African region has been limited by a combination of factors which include the following: *corruption; poor institutional framework and infrastructure; inadequate DREs planning policies; uncoordinated actions in the energy sector; pricing distortions which have placed renewable energy at a disadvantage, in particular the strong subsidy of fossil energies; high initial capital costs of DREs; weak dissemination strategies; poor decentralised solutions for energy services; lack of consumer awareness on benefits and opportunities of renewable energy solutions; unavailability of funds for development of renewable energies; lack of skilled manpower; poor baseline information; weak services and finally, weak or lack maintenance of infrastructures*.

Other authors [3] looked at the barriers from another perspective: the entrepreneurial setting. What constraints do Renewable Energy Entrepreneurs (REEs) in developing countries encounter while introducing DREs. Seven constraints were identified as key to REEs' success (or, conversely, failure) in developing countries: *inadequate or inappropriate government or policy support*, *inadequate local demand*, *price of DRESs*, *inadequate access to institutional finance*, *lack of skilled labour*, *underdeveloped physical infrastructure and logistics* and *power of incumbents* (*existing players on the energy market*).

Additionally, Yaqoot et al. [14] looked after decentralised renewable energy systems in more general, such as solar lanterns, solar home systems, family-type biogas plants, improved biomass cook stoves, etc. *Inappropriateness of technology, unavailability of skilled manpower for maintenance, unavailability of spare parts, high cost, lack of access to credit, poor purchasing power and other spending priorities, unfair energy pricing, lack of information or awareness and lack of adequate training on operation and maintenance of decentralised renewable energy systems* were found to be the most critical barriers [14]. The identified barriers have been classified under five broad categories depending on the characteristics of the barrier: technical, economic, institutional, sociocultural and environmental (see Table 2.1).

Table 2.1 Classification of barriers to the diffusion of DRE

Barrier	Sub-barriers
Technical	Resource availability; technology (design, installation and performance); skill requirement for design and development, manufacturing, installation, operation and maintenance
Economic	Cost; market structure; energy pricing; incentives; purchasing power and spending priorities; financial issues; awareness and risk perception
Institutional	Policy and regulatory; infrastructure (institutions for research, design and aftersales services); administrative
Socio-cultural	Societal structure; norms and value system; awareness and risk perception; behavioural or lifestyle issues
Environmental	Resources (land and water); pollution; aesthetics

Source Yaqoot et al. [14]

In conclusion, next to the opportunities for DRE in emerging markets, there are also a wide range of potential barriers. These barriers might vary per DRE technology, per region and per stakeholder perspective. For a successful implementation of DREs, it is critical to take these barriers in mind and to come up with remedial measures to overcome them. The used literature for this section can help to provide a deeper insight into the barriers as well as solutions to overcome them.

References

1. Beck F, Martinot E (2004) Renewable energy policies and barriers. In: Cleveland CJ (ed) Encyclopedia of energy. Academic Press, Elsevier Science, New York
2. Colombo E, Bologna S, Masera D (2013) Renewable energy for unleashing sustainable development. Springer, United Kingdom
3. Gabriel CA, Kirkwood J, Walton S, Rose EL (2016) How do developing country constraints affect renewable energy entrepreneurs? Rev Energy Sustain Develop 35(Supplement C):52–66. https://doi.org/10.1016/j.esd.2016.09.006
4. Karakaya E, Sriwannawit P (2015) Barriers to the adoption of photovoltaic systems: the state of the art. Rev Renew Sustain Energy Rev 49(Supplement C):60–6. https://doi.org/10.1016/j.rser.2015.04.058
5. Lambert J, Hall C, Balogh S, Poisson A, Gupta A (2013) EROI of global energy resources: status, trends and social implications. United Kingdom Department for International Development
6. Mackay DJC (2009) Sustainable energy—without the hot air. UIT Cambridge, UK
7. Mboumboue E, Njomo D (2016) Potential contribution of renewables to the improvement of living conditions of poor rural households in developing countries: Cameroon's case study. Rev Renew Sustain Energy Rev 61(Supplement C):266–79. https://doi.org/10.1016/j.rser.2016.04.003
8. National Renewable Energy Laboratory (NREL) (2012) Renewable electricity futures study. In: Hand MM, Baldwin S, DeMeo E, Reilly JM, Mai T, Arent D, Porro G, Meshek M, Sandor D (eds), 4 vols. NREL/TP-6A20-52409. National Renewable Energy Laboratory, Golden, CO

9. National Wind Coordinating Committee (NWCC) (2010) Wind turbine interactions with birds, bats, and their habitats: a summary of research results and priority questions
10. Rifkin J (2011) The third industrial revolution. In: How lateral power is transforming energy, the economy, and the world. P. Macmillan, New York
11. Rolland S, Glania G (2011) Hybrid mini-grids for rural electrification: lesson learned. Alliance for Rural Electrification, Brussels
12. Skierka K (2016) SDGs and DRE: the critical role of national policy in accelerating DRE markets and achieving power for all. In: Vezzoli C, Delfino E (eds) Sustainable energy for all by design conference proceedings, pp 23–31
13. World Energy Council (2004) Comparison of energy systems using life cycle assessment. World Energy Council, London (UK)
14. Yaqoot M, Diwan P, Kandpal TC (2016) Review of barriers to the dissemination of decentralized renewable energy systems. Rev Renew Sustain Energy Rev 58(Supplement C):477–90. https://doi.org/10.1016/j.rser.2015.12.224

Chapter 3
Sustainable Product-Service System (S.PSS)

3.1 S.PSS: An Introduction and Definition

A key contemporary query is the following: within the current social, environmental and economic crisis, which are the opportunities for innovate towards sustainability? Do we know any offer/business model capable of creating (new) value, decoupling it from material and energy consumption? In other words, is there any alternative to significantly reduce the environmental impact of traditional production/consumption systems?

One promising alternative is the development and implementation of sustainable product-service systems, which can be defined as an '***…offer model providing an integrated mix of products and services that are together able to fulfil a particular customer demand (to deliver a "unit of satisfaction"), based on innovative interactions between the stakeholders of the value production system (satisfaction system), where the ownership of the product/s and/or its life cycle responsibilities remain by the provider/s, so that the economic interest of the providers continuously seek new environmentally and/or socioethically beneficial solutions***' (adapted from Vezzoli et al. [22]).

Sustainable Product-Service System (S.PSS) has been studied since the end of the 90s as (one of) the most promising offer/business models in this perspective [3, 4, 8, 9, 12, 16, 24]. More recently, they demonstrate to be one of the most promising offer models to extend the access to goods and services even to low- and middle-income contexts, thus enhancing social equity and cohesion. S.PSS is understood as a win-win offer model combining the three pillars of sustainability, the economic with the environmental and the socioethical ones.

In fact, Sustainable Product-Service System (S.PSS) is value propositions introducing relevant innovation on different levels (see even Fig. 3.1).

- They shift the business focus from selling (only) products to offering a so-called 'unit of satisfaction', i.e. a combination of products and services jointly capable of achieving a final user satisfaction;

C. Vezzoli et al., *Designing Sustainable Energy for All*,
Green Energy and Technology, https://doi.org/10.1007/978-3-319-70223-0_3

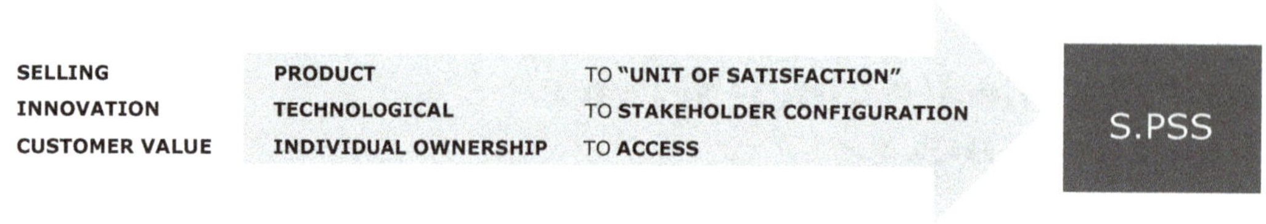

Fig. 3.1 S.PSS: a paradigm shift from traditional product offer. *Source* designed by the Authors

- They shift the primary innovation from a technological one to an innovation on a stakeholder interaction level, i.e. they are based on three main types of innovative stakeholder configurations: product offer combined with product life cycle services to customer, offer as enabling platform for customers and final results offer to customers;
- They shift the value perceived by the customer from individual ownership to access to goods and services.

Finally, as the key understanding of our discourse, S.PSSs are offer models with a win-win sustainability potential, i.e. they are offer/business models capable of creating (new) value decoupling it from resources consumption and environmental impact increase while extending access to goods and services to low- and middle-income people enhancing social equity and cohesion.

3.2 S.PSS Types

There is a continuum of approaches for an S.PSS configuration on which it is possible to identify three major S.PSS types to system innovation, which have been studied and listed as favourable to achieve higher levels of eco-efficiency [19, 21, 22].

1. Product-oriented S.PSS: services providing added value to the product life cycle;
2. Use-oriented S.PSS: services providing 'enabling platforms for customers';
3. Result-oriented S.PSS: services providing 'final results' for customers.

3.2.1 Product-Oriented S.PSS: Adding Value to the Product Life Cycle (Type I)

Let us start with an example of an eco-efficient system innovation adding value to the product life cycle.

Wilkhahn aftersale services for chairs.

During office swivel chairs life, periodical checks are carried out to keep the products in good working order. The order includes a service agreement which comprises three visits by service technicians within a period of 5 years. Older products, which no longer meet current technical or design standards, may be updated if the customer wishes. The customer can find the information about these opportunities on the product web-site. At the end of product life customers are offered take-back and recycling services. For furniture ranges, that are no longer produced, an additional repair service for two years is offered. A general overhaul is usually carried out at producer's plant based on a detailed estimate, and is arranged by the company consultant or by a local dealer. The producer company, guarantees the take-back of worn out products. They are disassembled, all parts are sorted into pure material categories and passed on for recycling. In the case of a new order, no take-back costs will be billed for those chairs being replaced by new chairs ordered from Wilkhahn. Wilkhahn interests do not rely only on the number of chairs sold, but also on service; in fact, the services provided help to reduce the number of produces to be entirely replaced. Clients perceive added value from the offered services because they free them from the costs and the problems associated with the monitoring and checking of their chairs. Achieving better efficiency from chairs and chair-services also provides many economic benefits both in production processes and in improving the life of chairs.

A **product-oriented S.PSS** innovation adding value to the product life cycle is defined as follows:

A company (alliance of companies) that provides additional services to guarantee an extended life cycle performance of the product/semi-finished product (sold to the customer).

A typical service contract would include maintenance, repair, upgrading, substitution and product take-back services over a specified period.

This reduces the user's responsibility in the use and/or disposal of the product/semi-finished product (owned by her/him), and the innovative interaction between the company and the customer drives the company's economic and competitive interest in continuously seeking environmentally beneficial new solutions, i.e. the economic interest becomes something other than only selling a larger number of products.

3.2.2 Use-Oriented S.PSS: Offering Enabling Platforms for Customers (Type II)

The following box describes an example of an eco-efficient S.PSS innovation as an enabling platform for customers.

Car sharing—Move About by Th!nk

Move About, like many other car-sharing systems, is a service providing an enabling platform of product (car) and services. It is a car-sharing scheme for the general public in Oslo; the fleet of vehicles is made up of 40 electric cars, all from the Norwegian manufacturer Th!nk. Users pay a monthly membership fee plus an hourly rate (including everything from the insurance to the energy to move the vehicle). For car users, a subscription to a car-sharing system provides convenient access to car mobility at lower costs than a traditional car rental agency. The local administration offers various incentives, such as free parking, exemption from road pricing and authorization to drive in bus lanes.[1] A car-sharing system intensifies the use of cars, meaning a lower number of cars are needed in a given context for a given demand for mobility.

A **use-oriented S.PSS** innovation offering as an enabling platform to customers is defined as follows:

A company (alliance of companies) offering access to products, tools, opportunities or capabilities that enable customers to meet the particular satisfaction they want (in other words efficiently satisfying a particular need and/or desire). The customer obtains the desired utility but does not own the product that provides it and pays only for the time the product is actually used.

Depending on the contract agreement, the user could have the right to hold the product/s for a given period of time (several continuous uses) or only for one use. Commercial structures for providing such services include leasing, pooling or sharing of certain goods for a specific use.

The client thus does not own the products and does operate them to obtain the final satisfaction (the client pays the company to provide the agreed results). Again, in this case, the innovative interaction between the company and the client drives the company's economic and competitive interest to continuously seek environmentally beneficial new solutions, e.g. to design highly efficient, long-lasting, reusable and recyclable products.

3.2.3 Result-Oriented S.PSS: Offering Final Results to Customers (Type III)

The following describes an example of an eco-efficient S.PSS innovation providing final results to customers.

Phillips, pay per lux service.

The 'pay per lux' is a full-service providing a final result, consisting of 'selling' light as a finished product. Light is delivered through a led system, which is produced and managed during its life by Phillips. Business customers pay a regular fee to Phillips that covers their entire lighting service – design, equipment, installation, maintenance and upgrades – only paying the 'lux', the light consumed. The innovation of this product-service system is that Phillips will not invoice the client for the energy consumed to obtain the 'lux', but rather, 'lux' is sold as an entire service. By planning for longevity rather than a with a product-sale approach, it provides the most efficient and cheapest lighting possible, thus encouraging the uptake of energy-saving lighting. At the end of the contract, products can be returned to the production process again, reusing the raw materials, optimising recycling and reducing waste.

[1] See www.mindsinmotion.net/index.php/mimv34/themes/hybrid_electric/featured/move_about.

A **result-oriented S.PSS innovation** offering final results to customers can be defined as follows:

A company (alliance of companies) that provides a customised mix of services (as a substitute for the purchase and use of products), in order to provide an integrated solution to meet a particular customer's satisfaction (in other words a specific final result). The mix of services does not require the client to assume (full) responsibility for the acquisition of the product involved. Thus, the producer maintains the ownership of the products and is paid by the client only for providing the agreed results.

The customer does not own the products and does not operate them to achieve the final satisfaction; the client pays the company to provide the agreed results. The customer benefits by being freed from the problems and costs involved in the acquisition, use and maintenance of equipment and products. The innovative interaction between the company and the client drives the company's economic and competitive interest to continuously seek environmentally beneficial new solutions, e.g. long-lasting, reusable and recyclable products.

Moreover, if properly conceived, S.PSS can offer to low- and middle-income people the opportunity to get access to services that traditional product sales models would not allow.

In fact, it has been argued that in low- and middle-income contexts 'a S.PSS innovation may act as a business opportunity to facilitate the process of a socio-economic development by jumping over the stage characterised by individual consumption/ownership of mass-produced goods—towards a 'satisfaction-based' and 'low resource-intensity' advanced service-economy' [20].

3.3 S.PSS Sustainability Benefits

The next paragraphs describe in detail the sustainability win-win potentials of S.PSS models in terms of environmental, socioethical and economic benefits.

3.3.1 S.PSS Environmental Benefits

When is an S.PSS eco-efficient? When can we decouple the economic interests from resource consumption and environmental impact in general? In other terms, why and when is an S.PSS producer/provider economically interested in design for environmental sustainability?

The following S.PSS environmental benefits (eco-efficient potentials) could be highlighted.

(a) As far as the S.PSS model is offering the products/s, retaining the ownership and being paid per unit of satisfaction, or offering all-inclusive the product with its maintenance, repair and substitution, the LONGER the product/s or its components last (*environmental benefits*), the MORE the producer/provider avoids/postpones the disposal costs plus the costs of pre-production, production and distribution of a new product substituting the one disposed of (*economic benefits*). Hence, the producer/provider is driven by economic interests to **design (offer) for lifespan extension of product/s** (*eco-efficient product LCD implications*).

(b) As far as the S.PSS model is selling a shared use of product/s (or some product's components) to various users, the MORE intensively the product/s (or some product's components) are used, i.e. being used most of the time (*environmental benefits*), the HIGHER the profit, i.e. proportionally to the overall use time (*economic benefits*). Hence, the producer/provider is driven by economic interests to **design for intensive use of product/s** (*eco-efficient product LCD implications*).

(c) As far as the S.PSS model is selling all-inclusive the access to products/s and the resources it consumes in the use phase, with payment based on unit of satisfaction (product's ownership by the producer/provider), the HIGHER the product/s resource efficiency in the use phase (*environmental benefits*), the HIGHER the profit, i.e. the payment minus (among others) the costs of resources (*economic benefits*). Hence, the producer/provider is driven by economic interests to **design/offer product/s that minimise resources consumption in the use phase** (*eco-efficient product LCD implications*).

(d) As far as the S.PSS model is selling energy as all-inclusive access to the energy production unit and the source for energy generation, with pay per period/time/satisfaction (energy production unit ownership by the producer/supplier), the HIGHER the use of passive/renewable sources of energy (*environmental benefits*), the HIGHER the profit, i.e. the payment minus (among others) the costs of non-passive/renewable sources of energy supplied (*economic benefits*). Hence, the producer/provider is driven by economic interests to **design (offer) for passive/renewable resources optimization** (*eco-efficient product LCD implications*).

(e) As far as the S.PSS model is selling all-inclusive the product with its end-of-life treatment/s, the MORE the materials are either recycled, incinerated with energy recovery, or composted (*environmental benefits*), the MORE are the avoided costs of landfilling and new primary material, energy or compost (*economic benefits*). Hence, the producer/provider is driven by economic interests to **design for material life extension (recycling, energy recovery or composting)** (*eco-efficient product LCD implications*).

(f) As far as the S.PSS model is selling all-inclusive the toxic or harmful product/s with use and/or end-of-life toxicity/harmfulness management services, the LOWER the potential toxic or harmful emissions during use and/or at the end-of-life (*environmental benefits*), the MORE the avoided costs of both toxic/harmful treatments in use and/or at the end-of-life. Hence, the producer/provider is driven by economic interests to **design (offer) for toxicity/harmfulness minimization** (*eco-efficient product LCD implications*).

To conclude, when is an S.PSS eco-efficient? When the product ownership and/or the economic responsibility of its life cycle performance remains by the producer/providers who are selling a unit of satisfaction rather than (only) the product.

And why does this happen? Because this way, we shift/allocate responsibility for the products and/or the services design/development, to the producers/providers, that in this way has direct economic and competitive interest in reducing the environmental impacts of their products/services.

Finally, within an S.PSS model, a product LCD/eco-design is eco-efficient. In other terms, an S.PSS producer/provider is economically interested in design for:

- Product lifespan extension and use intensification;
- Material life extension (recycling, energy recovery, composting);
- Resource (materials and energy) minimisation;
- Resource (materials and energy) renewability and biocompatibility;
- Resource (materials and energy) toxicity/harmfulness minimisation.

3.3.2 S.PSS Socioethical Benefits

Why S.PSS may foster socioethical benefits? Because S.PSS make goods and services accessible to both final users and entrepreneurs even in low- and middle-income contexts. The following S.PSS socioethical benefits (social equity and cohesion potentials) could be highlighted.

(a) As far as the S.PSS model is selling the access rather than mere product ownership, this reduces/avoids purchasing costs of products which are frequently too high for low- and middle-income people *(economic benefits),* i.e. making goods and services more easily accessible (*socioethical benefits*).

(b) As far as the S.PSS model is selling the 'unit of satisfaction' including life cycle services costs, this reduces/avoids running cost for maintenance, repair, upgrade, etc. too high for low- and middle-income people *(economic benefits)*, i.e. avoiding interruption of product use (*socioethical benefits*).

(c) As far as the S.PSS model is selling access rather than working equipment, this reduces/avoids initial (capital) investment costs of equipment, frequently too high for low- and middle-income entrepreneurs *(economic benefits)*, i.e. facilitating new business start-up in low- and middle-income contexts (*socioethical benefits*).

(d) As far as the S.PSS model is selling entrepreneurs all-inclusive life cycle services with the equipment offer, this reduces/avoids running cost for equipment maintenance, repair, upgrade, etc. frequently too high for low- and middle-income entrepreneurs *(economic benefits),* i.e. this avoids interruption of equipment use (working activities) (*socioethical benefits*).

(e) As far as the S.PSS model is offering goods and services without purchasing costs, this opens new market opportunities for local entrepreneurs via new potential low- and middle-income customers (BoP), i.e. potentially empowering locally based economies and improving quality of life (*socioethical benefits*).

3.3.3 S.PSS Economic and Competitive Benefits

What are the main economic and competitive benefits of S.PSS? The following S.PSS economic and competitive benefits could be highlighted:

(a) As far as the S.PSS model offer service along all its life cycle, they can establish longer and stronger relationships with customers, i.e. increasing customer fidelity.
(b) As far as the S.PSS model is different from traditional product sales which are nowadays in saturated market, they can open up new business opportunities, i.e. empowering strategic positioning.

3.4 S.PSS Barriers and Limits

3.4.1 Not All PSSs Are Sustainable

It is important to underline that not all shifts to PSS result in environmental benefits: a PSS must be specifically designed, developed and delivered, if it is to be highly eco-efficient. For example, schemes where products are borrowed and returned incur transportation costs (and the resultant use of fuel as well as polluting emissions) over the life of the product. In some specific instances, the total fuel cost and environmental impact may make the system non-viable in the long term.

Furthermore, even when well designed, it has been observed that some PSS changes could generate unwanted side effects, usually referred to as rebound effects.

Society is a set of complexes, interrelated systems that are not clearly understood. As a result, something may happen that turns potential environmentally sound solutions into an increase in global consumption of environmental resources at the practical level. One example is the impact of PSS on consumer behaviour. For example, outsourcing, rather than ownership of products, could lead to careless (less ecological) behaviours.

Nevertheless, S.PSS development seen presents great potential for generating win-win solutions that promote profit and environmental benefits. It has the potential to provide the necessary, if not sufficient, conditions to enable communities to leapfrog to less resource-intensive (more dematerialised) systems of social and economic systems.

3.4.2 Barriers

Barriers to overcome may include a lack of external infrastructure and technologies, e.g. for product collection, remanufacturing or recycling. Per stakeholder type, barriers for the eco-efficient PSS diffusion in industrialised contexts are summarised as follows [5, 6]:

- For **companies**, the adoption of an S.PSS strategy is more complex to be managed than the existing way of delivering products alone. There is a need to implement changes in corporate culture and organisation to support a more systemic innovation and service-oriented business [20]; there is indeed resistance by companies to extend involvement with a product beyond point-of-sale [14, 18]. Extended involvement requires new design and management knowledge and approaches. It requires medium-to-long-term investments and is therefore connected with uncertainties about cash flows [16]. Moreover, a further obstacle is the difficulty of quantifying the savings arising from S.PSS in economic and environmental terms, in order to market the innovation to stakeholders both inside and outside the company, or to the company's strategic partners [20]. Finally, the significant change in the system of earning profit could deter producers from employing the concept, first through limited experience in pricing such an offering, and second through fear of absorbing risks that were previously assumed by customers [1];
- For **customers/users**, the main barrier is the cultural shift necessary to value an ownerless way of having a satisfaction fulfilled, as opposed to owning a product [10, 13, 14, 20]. Solutions based on sharing and access contradict the dominant and well-established norm of ownership [2]; this is especially true in the B2C market, while in the B2B sector numerous examples of eco-efficient PSS concepts can be identified [17]. Product ownership not only provides a function to private users but also status, image and a sense of control [11]. Another obstacle is the lack of knowledge about life cycle costs [23], which makes it difficult for a user to understand the economic advantages of ownerless solutions;
- For **governments**, on the regulatory and policy side, actual laws may not favour S.PSS-oriented solutions. Environmental innovation is often not rewarded at the company level due to lack of internalisation of environmental impacts [15]. In addition, there are difficulties in implementing policies to create corporate drivers to facilitate the promotion and diffusion of this kind of innovation [7, 15].

References

1. Baines T, Lightfoot H, Evans S, Neely A, Greenough R, Peppard J, Roy R, Shehab E, Braganza A, Tiwari A, Alcock J, Angus J, Bastl M, Cousens A, Irving P, Johnson M, Kingston J, Lockett H, Martinez V, Micheli P, Tranfield D, Walton I, Wilson H (2007) State-of-the-art in product service-systems. Proc Inst Mech Eng Part B J Eng Manuf 221 (10):1543–1552
2. Behrendt S, Jasch C, Kortman J, Hrauda G, Pfitzner R, Velte D (2003) Eco-service development: reinventing supply and demand in the European Union. Greenleaf Publishing, Sheffield
3. Bijma A, Stuts S, Silvester S (2001) Developing eco-efficient product-service combinations. In: Proceedings of the 6th international conference sustainable services and systems: transition towards sustainability? Amsterdam, The Netherlands, October 2001

4. Brezet JC, Bijma AS, Ehrenfeld J, Silvester S (2001) The design of eco-efficient services: method, tools and review of the case study based designing eco-efficient services. Dutch Ministries of Environment VROM, Delft University of Technology, Delft, The Netherlands
5. Ceschin F (2013) Critical factors for implementing and diffusing sustainable product-Service systems: insights from innovation studies and companies experiences. J Cleaner Prod 45:74–88. ISSN: 0959-6526
6. Ceschin F (2014) Sustainable product-service systems: between strategic design and transition studies. Springer International Publishing. ISSN 13: 978-3-319-03794-3
7. Ceschin F, Vezzoli C (2010) The role of public policy in stimulating radical environmental impact reduction in the automotive sector: the need to focus on product-service system innovation. Int J Automot Technol Manage 10(2/3):321–341
8. Charter M, Tischner U (2001) Sustainable solutions: developing products and services for the future. Greenleaf Publishing, Sheffield, UK
9. Cooper T, Sian E (2000) Products to Services. Friends of the Earth, Centre for Sustainable Consumption, Sheffield Hallam University
10. Goedkoop M, van Halen C, Te Riele H, Rommes P (1999) Product services systems, ecological and economic basics, report 1999/36. VROM, The Hague
11. James P, Hopkinson P (2002) Service innovation for sustainability. A new option for UK environmental policy?. Bradford University, Bradford
12. Manzini E, Vezzoli C (2001) Strategic design for sustainability. TSPD proceedings, Amsterdam
13. Manzini E, Vezzoli C, Clark G (2001) Product service systems: using an existing concept as a new approach to sustainability. J Design Res 1(2)
14. Mont O (2002) Clarifying the concept of product–service system. J Clean Prod 10:237–245
15. Mont O, Lindhqvist T (2003) The role of public policy in advancement of product service systems. J Clean Prod 11(8):905–914
16. Mont O (2004) Product-service systems: panacea or myth? Ph.D. Dissertation. IIIEE, University of Lund, Sweden
17. Stahel WR (1997) The functional economy: cultural and organizational change. From the industrial green game: implications for environmental design and management. National Academy Press, Washington (DC)
18. Stoughton M, Shapiro KG, Feng L, Reiskin E (1998) The business case for EPR: a feasibility study for developing a decision-support tool. Tellus Institute, Boston
19. Tukker A (2004) Eight types of product-service system: eight ways to sustainability? Experiences from SusProNet. Bus Strategy Environ 13:246–260
20. United Nations Environment Programme (UNEP) (2002) Product-service systems and sustainability. Opportunities for sustainable solutions. UNEP, Paris
21. United Nations Environmental Program (UNEP) (2009) Design for sustainability. A step-by-step approach. UNEP, Paris
22. Vezzoli C, Kohtala C, Srinivasan A, Xin L, Fusakul M, Sateesh D, Diehl JC (2014) Product-service system design for sustainability. Greenleaf Publishing Inc, London
23. White AL, Stoughton M, Feng L (1999) Servicizing: the quiet transition to extended product responsibility. Tellus Institute, Boston
24. Zaring O, Bartolomeo M, Eder P, Hopkinson P, Groenewegen P, James P, de Jong P, Nijhuis L, Scholl G, Slob A, Örninge M (2001) Creating eco-efficient producer services. Gothenburg Research Institute, Gothenburg

Chapter 4
Sustainable Product-Service System Applied to Distributed Renewable Energies

4.1 Sustainable Product-Service System Applied to Distributed Renewable Energy: A Win-Win Opportunity

We argued in previous chapters that Distributed Renewable Energy (DRE) generation is a promising approach towards sustainable energy for All. Aside, we described the Sustainable Product-Service System (S.PSS) model, as promising one towards sustainable development, even in low- and middle-income contexts. In this chapter, we describe the application of the Sustainable Product-Service System (S.PSS) to Distributed Renewable Energy (DRE) as a win-win opportunity for the diffusion of sustainable energy, even in low- and middle-income contexts.

It is clear that we need to undergo a paradigm shift in the way we produce, supply and use the energy.

Indeed, to reach the shift, will by coupling the two models, mean to: shift from centralised and non-renewable energy system to distributed renewable energy systems, in which the user can be the prosumer (consumer + producer) of her/his energy with small generator units nearby or at the point of use sourced by sun, wind and all other forms of renewable energy. Furthermore, in case of energy systems, the shift from individual ownership consumption to Sustainable Product-Service System would entail that:

- would be a model where the providers retain the ownership or at least some responsibilities over the life cycle of the small generator unit (of renewable energy) and eventually of the products that use the electricity, i.e. the Energy-Using Products/Equipment (EUP, EUE);
- customers pay per use/period, thus reducing/avoiding the (initial) investment cost of the small energy generator unit and eventually of the Energy-Using Products/Equipment;

C. Vezzoli et al., *Designing Sustainable Energy for All*,
Green Energy and Technology, https://doi.org/10.1007/978-3-319-70223-0_4

- the shift from the individual ownership to the satisfaction of an (energy) need, leading to avoid (unexpected) life cycle costs for the customer, as maintenance or repair on the small energy generator unit and eventually of the Energy-Using Products/Equipment, thus reducing the risk of drop-off.

In the next paragraph, a scenario of this paradigm shift is presented (Fig. 4.1).

4.2 Scenario for S.PSS Applied to Distributed Renewable Energy

This chapter describes the scenario[1]—a new picture and the new narration of sustainable production and consumption systems—characterised by the application of the promising model of Sustainable Product-Service Systems (S.PSS) to Distributed Renewable Energy (DRE) in low- and middle-income contexts, and aimed to inspire the design of sustainable energy solutions, accessible by All.

This scenario is described by four visions, each representing a possible win-win configuration of S.PSS applied to DRE in low- and middle-income contexts, i.e. combining sociocultural, organisational and technological factors, fostering solutions with a low environmental impact, a high socioethical quality and a high economic and competitive value.

The four visions narrate the scenario and are outlined within two polarity axes. The horizontal axis, i.e. different customers of the sustainable energy solutions, the final user (B2C—either individual or local community), or entrepreneur/business (B2B). The vertical axis highlights whether the energy solution offers Distributed Renewable Energy generator (e.g. solar panel system plus its components such as storage, inverter, wires), or both the Distributed Renewable Energy generator and one or more Energy-Using Products or Energy-Using Equipments (e.g. phone and television are Energy-Using Products; woodworking machine and sewing machine are Energy-Using Equipment).

The following narration of the four visions, which emerged as an intersection of the two axes, defines the picture of the overall scenario:

- *Energy for all in daily life* (Vision 1);
- *Energise your business without initial investment cost* (Vision 2);
- *'Pay x use' for your daily life products and energy* (Vision 3) and
- *Start-up your business paying per period for equipment and energy* (Vision 4) (Fig. 4.2).

Below a short description introduces each vision of the scenario (Fig. 4.3).

[1]A Sustainability Design Orienting Scenario (SDOS) has been developed as the application of the Sustainable Product-Service System (S.PSS) model to Distributed Renewable Energy (DRE). The scenario was developed with the following steps: case studies research, guidelines definition, workshop sessions and visions development.

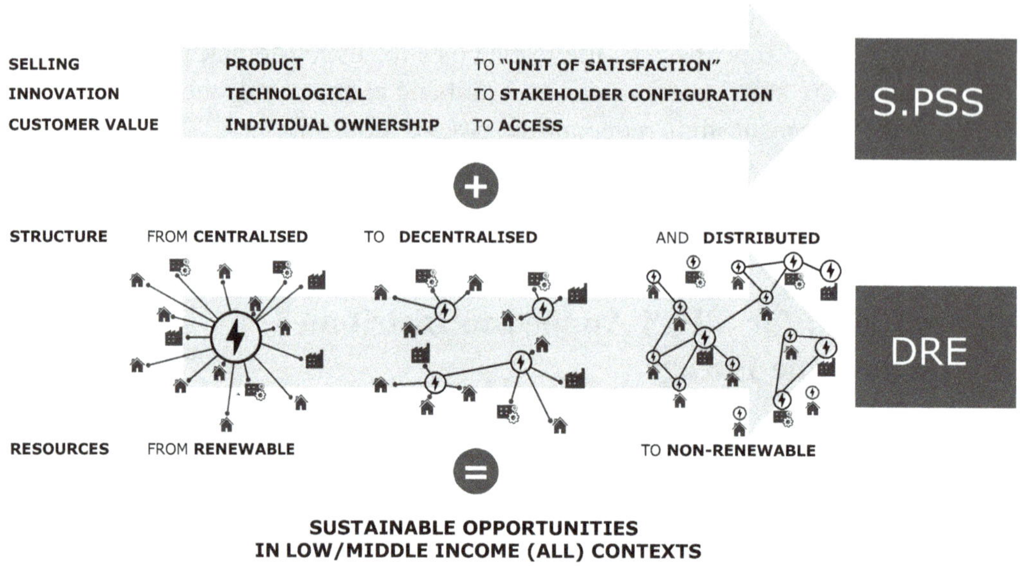

Fig. 4.1 The coupling of the 2 paradigm shifts represented by S.PSS and DRE. *Source* designed by the Authors

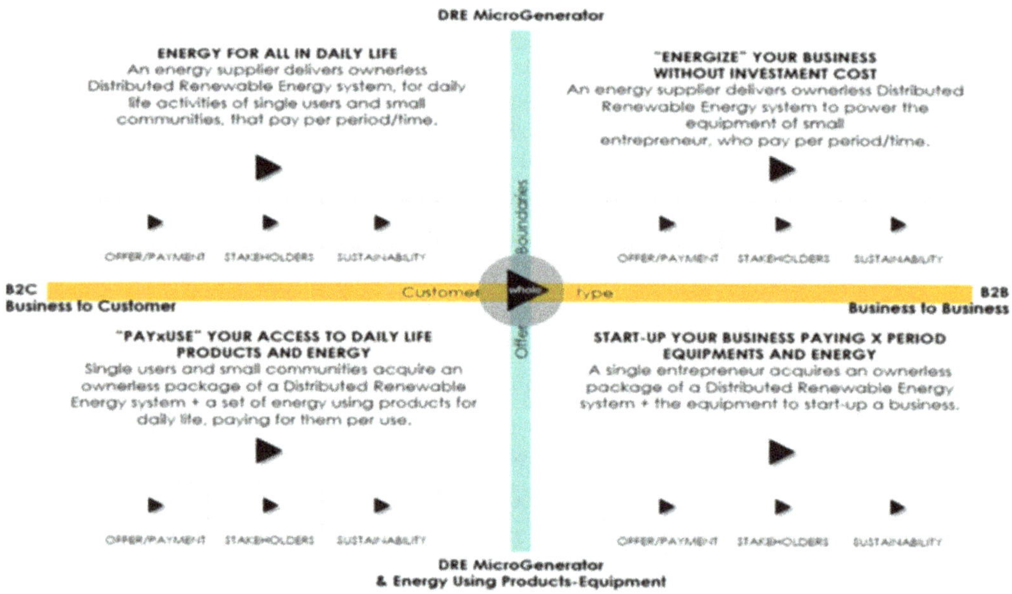

Fig. 4.2 The sustainable energy for all scenario. *Source* designed by the Authors

4.2.1 *Energy for All in Daily Life (Vision 1)*

The *energy for all in daily life* presents the offer of a Distributed Renewable Energy (DRE) micro-generator to a final customer (B2C). Indeed, it could be that: '*an energy supplier delivers an ownerless Distributed Renewable Energy system, for daily life activities, to single users and small communities who pay per period/*

Fig. 4.3 Screenshots from the video 'Energy for all in daily life'. *Source* Vanitkoopalangkul [58]

time'. This implies that the ownership of the Distributed Renewable Energy system (e.g. solar panel, wires, storage) stays with the energy supplier, who covers both the initial investment cost (e.g. the purchase of micro-generator and its components and their installation) and life cycle costs (e.g. maintenance and repair). The customer makes customisable periodic payments to access his/her (energy) satisfaction. This configuration makes access to energy economically affordable even in low- and middle-income contexts, so that the quality of life could be greatly improved, especially in relation to health and security.

The following short story illustrates one possible situation in a low-income context: '*Max, inhabitant of a rural village, has no access to energy. Therefore, he uses an oil lamp for light and he goes to the closest village to charge his phone. If he can have a solar system installed on his roof, guaranteeing secure energy access, he can avoid daily problems, and improve his and his family's quality of life.*'

The story could change coherently with the *energy for all in daily life* vision presented above (many other could be imagined): '*Max doesn't have to buy the Distributed Renewable Energy micro-generator and its components; he just uses them by paying per period a fixed amount of money. Ownership and related services remain with the energy supplier, who is interested in reducing maintenance and repair needs, improving his own profit while reducing the environmental impact of the system*' (Fig. 4.4).

A video of this story is available at: https://www.youtube.com/watch?v=93NXZLpxnUQ.

Fig. 4.4 Screenshots from the video energise your business without investment cost. *Source* Vanitkoopalangkul [58]

4.2.2 Energise Your Business Without Initial Investment Cost (Vision 2)

The *energise your business without initial investment cost* introduces a business-to-business (B2B) opportunity, in which '*an energy supplier delivers an ownerless Distributed Renewable Energy system to power the equipment of a small entrepreneur, who pays per period/time*'. Even in this case, the Distributed Renewable Energy system is not owned by the customer. This reduces the risks for the customers, such as small entrepreneurs or businesses, who do not have to face any initial investment, except for the purchase of the necessary Energy-Using Equipment (e.g. sewing machine for a tailor shop) which are not included in the offer. In this way, small entrepreneurs/businesses even in low- and middle-income contexts can receive stable energy access, thus being able to guarantee the production/delivery of a predetermined quantity of products/services within a given time, thus satisfying clients and opening market opportunities.

The following short story illustrates one possible situation in a low-income context: '*Kate and Tom, tailors in a rural village, have no stable access to Energy consequently they still use a diesel generator to power their sewing machine. If they can have a solar system installed in their tailor shop, guaranteeing secure energy access, they can guarantee on time delivery and avoid losing clients.*'

The story could change coherently with *energise your business without initial investment cost* vision presented above (many other could be imagined): '*Kate and Tom don't have to buy the Distributed Renewable Energy micro-generator and components, but only have to pay a fixed rate per period. Ownership and related*

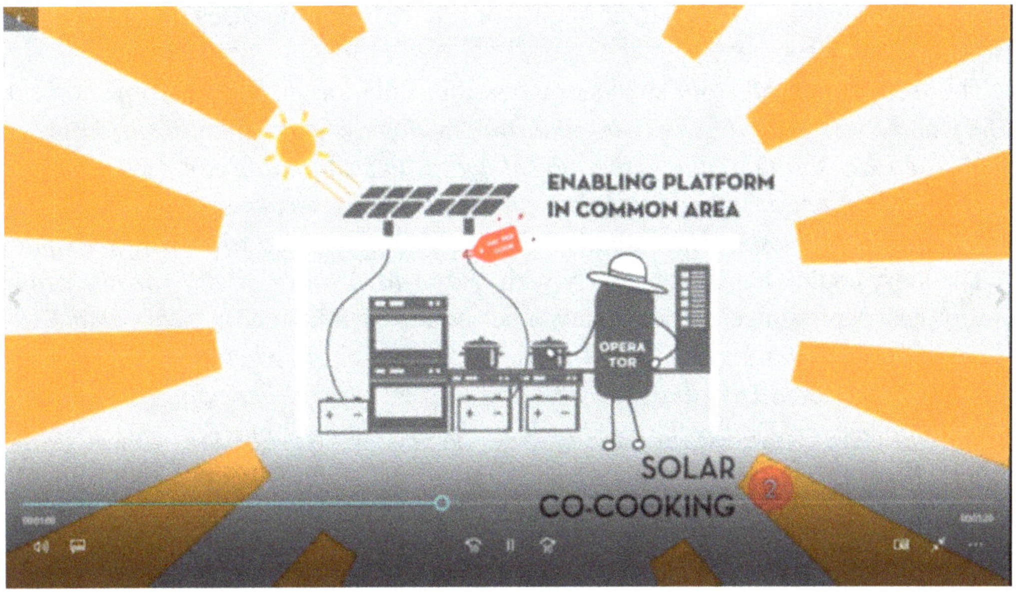

Fig. 4.5 Screenshots from the video 'Pay x use' for your daily life products and energy. *Source* Vanitkoopalangkul [58]

services stay with the energy supplier, who is interested in reducing maintenance and repair needs, improving his own business while reducing the environmental impact' (Fig. 4.5).

A video of this story is available at: https://www.youtube.com/watch?v=DB3XSYJ3wvg.

4.2.3 *'Pay x Use' Your Daily Life Products and Energy (Vision 3)*

The '*Pay x use*' *your daily life products and energy* presents a business-to-customer (B2C) offer of a Distributed Renewable Energy (DRE) micro-generator (and the related components) plus related Energy-Using Products, where '*single users and small communities acquire an ownerless package consisting of a Distributed Renewable Energy system plus a set of energy using products for daily life, paying for them per use.*' Similar to the two previous visions, even in this case, the Distributed Renewable Energy micro-generator and the related components are owned by the energy supplier. Different from previous visions, the Energy-Using Products (e.g. burner, oven) are included in the ownerless offer to the customer. This configuration cuts the initial investment costs (e.g. purchase, installation) of both Distributed Renewable Energy micro-generator and Energy-Using Products, as well as their life cycle costs (e.g. maintenance and repair) for the customers. The customers pay for the (energy) services they use, thus increasing affordability of the solution. For many people who still using firewood for cooking, access to clean energy could greatly improve their quality of life, reducing diseases caused by toxic emissions from the fire.

The following short story shows one possible situation in a low-income context: '*Mary and Ryan, are a family living in a rural village where cooking with firewood is still the main solution, due to the lack of access to sustainable energy. If they can have a solar system installed on their roof, guaranteeing secure energy access, they can reduce health risks, while gaining time no longer needed to collect firewood.*'

The story could change coherently with '*Pay x use*' *your daily life products and energy* vision presented above (many other could be imagined): '*Mary and Ryan can use the common kitchen based in the village to cook, where the energy used comes from the local Distributed Renewable Energy system. They don't have to buy any component or Energy Using Products in the kitchen, but they pay to cook. Ownership and related services stay with the energy supplier, who is interested in reducing maintenance and repair needs, improving their own business while reducing environmental impact.*'

A video of this story is available at: https://www.youtube.com/watch?v=ri5IPoIO_6Q.

4.2.4 Start-up Your Business Paying Per Period for Equipment and Energy (Vision 4)

The '*Start-up your business paying per period for equipment and energy*' presents an offer for small entrepreneurs/businesses (B2B) offer where '*a single entrepreneur acquires an ownerless package, consisting of a Distributed Renewable Energy system plus the equipment to start-up a business*'. In this configuration, a small entrepreneur/business receives an ownerless Distributed Renewable Energy system package (e.g. carpenter's workshop) composed of a Distributed Renewable Energy micro-generator and related components and the related Energy-Using Equipment (e.g. circular saw, drill). The ownership of the full package is retained by the energy supplier or a partnership. This cuts the initial investment costs for the purchase of both the Distributed Renewable Energy micro-generator and the Energy-Using Equipment, as well as their life cycle costs for the small entrepreneur/business. This comes to be very relevant, especially in low- and middle-income contexts, where many small entrepreneurs/businesses cannot get a loan from traditional banks. With stable access to energy, they could increase their business opportunities and working conditions, while empowering local economic growth.

The following story shows one possible situation in low-income context: '*Ben, carpenter in a big city, wants to move back to his own village to open a carpentry workshop, but no energy access is available. If he can have a solar system installed in his carpentry workshop in the village, guaranteeing secure energy access, he can start his business, offering on time delivery, with the most updated and energy efficient energy using equipment.*'

The story could change coherently with *start-up your business paying per period for equipment and energy* vision above (many other could be imagined): '*Ben*

Fig. 4.6 Screenshots from the video 'Start-up your Business' paying per period for equipment and energy. *Source* Vanitkoopalangkul [58]

doesn't have to buy the Distributed Renewable Energy micro-generator and components or the Energy Using Equipment for his shop. He just pays a fixed rate per period. Ownership and related services stay with the energy supplier who is interested in reducing maintenance and repair needs, improving their own business while reducing environmental impact' (Fig. 4.6).

A short video of this possible offer is available at: https://www.youtube.com/watch?v=14sdFSI8A5M.

4.3 S.PSS Applied to DRE: Sustainability Potential Benefits

The potential benefits of S.PSS and DRE have been extensively discussed for each model respectively on Sects. 3.3 and 2.1. This section discusses the potential sustainability benefits derived from applying S.PSS to DRE. It has, in fact, been argued that the combination of these two models represents a promising approach to deliver sustainable energy solutions in low- and middle-income contexts [14, 57]. Several potential advantages can be identified [17].

4.3.1 Environmental Benefits of S.PSS Applied to DRE

The adoption of a S.PSS approach in DRE solutions would make energy providers/manufacturers economically interested in seeking after environmentally beneficial solutions (as fully discussed in Sect. 3.3.1). In fact, if providers retain ownerships and responsibilities over the DRE generation unit/s and Energy-Using Products/ Equipment involved in the offer, the providers will be interested in extending lifespan of the physical elements of the solution, as well as interested in extending materials lifespan through recycling, energy recovery or composting. Furthermore, if providers are paid based on the performance delivered (and not per unit of product sold), they would be interested in reducing as much as possible the material and energy resources needed to provide that performance, as well as to design (offer) for passive/renewable resources optimisation. Additionally, in the case where S.PSS is applied to DRE solutions that promote a shared use of product/s (or some product's components, i.e. DRE generation unit, or Energy-Using Products/ Equipment) by multiple users, the more intensively these products are used, the higher the profit will be, i.e. proportional to the overall use time. So forth, having manufacturers/providers to keep ownership or at least some responsibilities over the DRE generator units and Energy-Using Products/Equipment, it represents a crucial aspect to encourage design for intensive use of product/s. Finally, if the S.PSS applied to DRE solution includes an all-inclusive service package to manage the toxic or harmful product/s in use and/or end-of-life, the producer/provider is driven by economic interests to design (offer) for toxicity/harmfulness minimisation.

4.3.2 Socioethical Benefits of S.PSS Applied to DRE

From a user's perspective, a S.PSS approach applied to DRE can increase customer (energy) satisfaction. This could happen because S.PSS offers access to (energy) satisfaction rather than mere (energy) product ownership, thus reducing/avoiding initial investment costs and running costs, e.g. maintenance and repair of energy products, which are frequently too high for low- and middle-income customers. In addition, a S.PSS applied to DRE offer can be tailored to the customers' particular (cultural and ethical) needs more easily than traditional product-based offers while making goods and (energy) services more easily accessible to All and increasing reliability.

From a business perspective, since S.PSSs are characterised by being labour and relationship-intensive solutions, and since both DRE and S.PSS require labour activities to be carried out at a local level, this can lead S.PSS applied to DRE solutions to a greater involvement of more local, rather than global, socio-economic stakeholders. This could result in an increase in local employment (as explained before) and local dissemination of skills and competences [53, 56], i.e. facilitating new business start-up in low- and middle-income contexts. Additionally, the selling of all-inclusive life cycle services with the equipment might reduce/avoid running cost for equipment maintenance, repair, upgrade, etc., frequently too high for low- and middle-income entrepreneurs, thus avoiding interruption of equipment use. Finally, S.PSS applied to DRE solutions offers (energy) services/business opportunities without initial investment costs, thus opening new market opportunities for local entrepreneurs via new potential low- and middle-income customers.

4.3.3 *Economic Benefits of S.PSS Applied to DRE*

From a user's perspective, S.PSS applied to DRE solutions does not require upfront payment for the products included (DRE generation units and potentially Energy-Using Products/Equipment). As result, low-income consumers can easily get access to modern electricity services without the need of making high initial investments [54].

From a business perspective, adopting a S.PSS approach can improve the strategic positioning and competitiveness of manufacturers/providers [21, 41, 62], establish a longer and stronger relationship with customers [11, 40] and build up barriers to entry for potential new competitors [21]. Finally, S.PSS applied to DRE solutions offers opportunities to strengthen the local economy and increase local employment. In fact, compared to traditional offers, S.PSS is more focused on the context of use, meaning that the service elements must be created at the same time and often at the same place when and where they are consumed [55]. Thus, skilled personnel might be empowered at a local level to carry out services such as installation, maintenance, repair, training, etc. The same is true for DRE systems which, compared to centralised systems, are characterised by a multiplicity of energy production units dispersed in the territory.

Combining a S.PSS approach offers additional advantages. From a user's perspective, a S.PSS approach can *increase customer satisfaction* because a S.PSS offer can be tailored to their particular (cultural and ethical) needs more easily than traditional product-based offers [53].

From a community angle, since S.PSSs are characterised by being labour and relationship-intensive solutions, and since both DRE and S.PSS require labour activities to be carried out at a local level, this can lead to a greater involvement of more local, rather than global, socio-economic stakeholders. Therefore, this could result in an *increase in local employment* (as explained before) *and local dissemination of skills and competences* [53, 56].

Of course, these potential benefits must be verified case-by-case, and balanced against the potential limitations and rebound effects (such as, for example, careless behaviours of users on not owned products). For this reason, S.PSS applied to DRE must be specifically designed, developed and delivered, in order to generate the above-mentioned sustainability advantages.

4.4 S.PSS Applied to DRE: A New Classification System and 15 Archetypal Models

S.PSS and DRE have been widely studied over the past decades, and knowledge has been built on how to classify these models. However, S.PSS and DRE have been only studied separately, and thus a comprehensive classification that looks at the combination of these two models is missing. This section puts forward a new classification system for S.PSS applied to DRE and describes 15 archetypal models of S.PSS applied to DRE [17].

4.4.1 Classification System

In the development of the new classification system, the starting point was the identification of the characterising dimensions used to classify S.PSS and DRE (Table 1.1).

Regarding DRE, several approaches have been proposed in the past to classify DRE models. These classification systems are built considering different combinations of characterising dimensions: *energy system*, including type of energy generation and type of energy source (e.g. [36]), *value proposition and payment structure* (e.g. [19]), *capital financing* (e.g. [4]), *energy system ownership* (e.g. [48]), *energy system operation* [50], *organisational form* (e.g. [64]) and *target customer* (e.g. [64]). It is important to highlight that no classification system encompasses all these characterising dimensions. They focus on a few (or sometimes only a single) dimensions, and thus they are individually unable to cover the complexity characterising DRE models.

Regarding S.PSS, the majority of the S.PSS classifications proposed in the past agree on three main S.PSS categories: product-oriented, use-oriented and result-oriented S.PSSs [52]. Gaiardelli et al. [20] carried out an extensive analysis on the dimensions taken into consideration in these classifications and identified five main characterising dimensions: *value proposition*, *product ownership*, *product operation*, *provider/customer relationship* and *environmental sustainability potential*.

As shown in Table 4.1, some of the identified characterising dimensions overlap, while some other are specifically used for DRE or S.PSS. For this reason, there is a need for a new classification system capable of simultaneously taking into consideration all the major dimensions characterising S.PSS applied to DRE (see Table 4.2).

Table 4.1 List of S.PSS and DRE dimensions

DRE dimensions	S.PSS dimensions
1. Energy system	–
2. Value proposition/payment structure	2. Value proposition/payment structure
3. Capital financing	–
4. Energy system ownership	4. Product ownership
5. Organisational form	–
6. Energy system operation	6. Product operation
7. Target customer	–
–	8. Provider/customer relationship
–	9. Environmental sustainability potential

Source Emili et al. [17]

Table 4.2 Dimensions characterising S.PSS applied to DRE

S.PSS&DRE Dimension	Description	Details
1. Energy system	Defines the connection type (stand-alone, grid-based systems) and renewable source involved (solar, wind, biomass etc.)	*Stand-alone system*: mini-kit, individual energy system, charging station *Grid-based system*: isolated mini-grid, connected mini-grid *Energy sources*: solar, hydro, biomass, wind, human power
2. Value proposition/ payment structure	Represents the value offered to the customer, i.e. the combination of product and services for which the customer is willing to pay and the payment structure	*Product-oriented S.PSS*: – Pay-to-purchase with advice, training and consultancy services – Pay-to-purchase with additional services *Use-oriented S.PSS*: – Pay-to-lease – Pay-to-share/rent/pool *Result-oriented S.PSS*: – Pay-per-energy consumed – Pay-per-unit of satisfaction
3. Capital financing	Indicates how capital is provided for the energy solution and determines cost recovering and tariff structure	Fully subsidised, quasi-commercial, commercially led
4. Ownership (of the energy system and Energy-Using Products)	Refers to who owns the energy system and products involved in the offer, i.e. the provider, the end user or a number of users	Customer or provider
5. Organisational form	Indicates the nature of the organisation providing the energy solution	Public sector-based, utility, NGO, community, PPP/hybrid, private sector-based
6. Energy system operation	Defines who operates the energy system	Customer or provider
7. Target customer	Indicates the type of end user (e.g. household, community, public building)	Individual customer or community
8. Provider/ customer relationship	Refers to the nature and intensity of interaction between the two actors and varies from transaction-based (product-oriented S.PSSs) to relationship-based (result-oriented S.PSSs) according to the responsibilities and activities performed on the product [20], [44]	Transaction-based or relationship-based
9. Environmental sustainability potential	Refers to the S.PSS environmental impact, which can potentially be lower than traditional product-based business models. It generally goes from high sustainability potential in result-oriented S.PSSs, to low sustainability potential in product-oriented S.PSSs [53]	Low, medium or high environmental sustainability potential

Source Emili et al. [17]

The new classification system was developed as a polarity diagram, in the attempt of grouping the major S.PSS&DRE characterising dimension into two groups (see Fig. 4.7).

The vertical axis includes:

- Value proposition (dimension #2);
- Ownership (of energy system and Energy-Using Products) (dimension #4);
- Energy system operation (dimension #6);
- Provider/customer relationship (dimension #8) and
- Environmental sustainability potential (dimension #9).

These dimensions can in fact overlap one another:

- The *value proposition* (dimension #2) ranges from product-oriented to use-oriented and result-oriented S.PSSs. This dimension is therefore strictly related to the *ownership* (of energy system and Energy-Using Products) (dimension #4). In fact, in product-oriented S.PSSs, the final user is the owner of the product/s, while moving towards result-oriented services the ownership is retained by the provider;
- The *value proposition* can also be aligned with the *energy system operation* dimension (#6), which refers to the management and operation of energy systems. In product-oriented and use-oriented S.PSSs, customers operate the energy

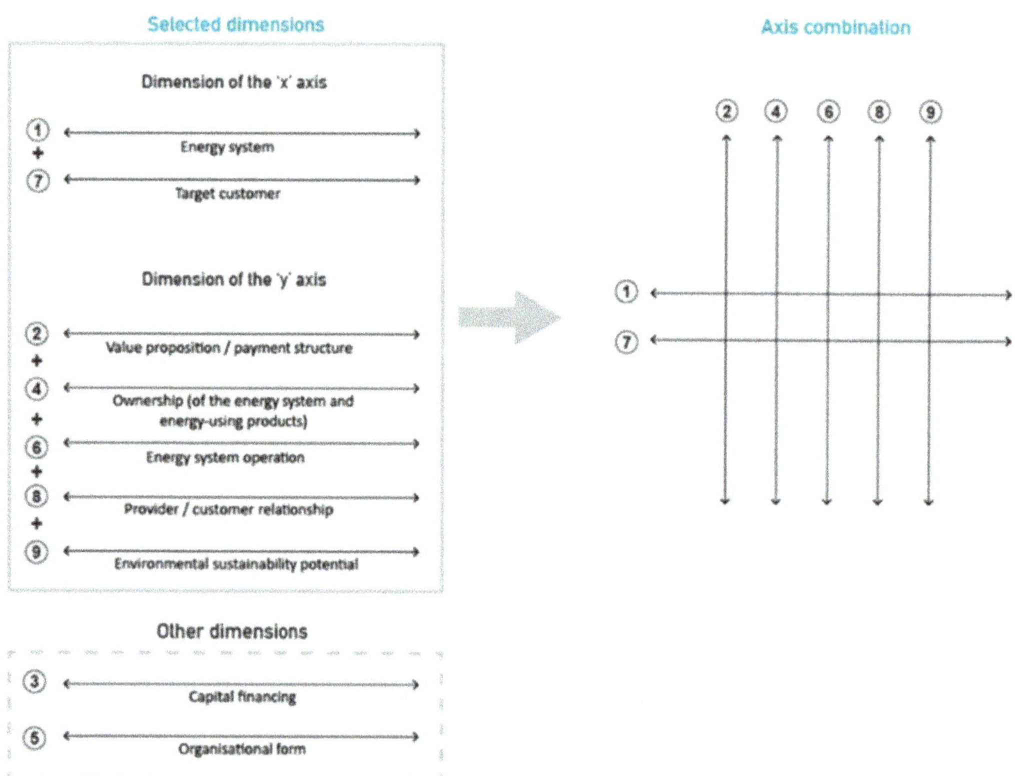

Fig. 4.7 Selection of dimensions' polarities and combination of axis used to build the classification system. *Source* Emili et al. [17]

systems to achieve the results they aim to. In result-oriented S.PSSs, the provider is responsible for operating the system in order to deliver the agreed final result to the customer. When Energy-Using Products are included in the offer, their operation is always performed by end users (e.g. using lamps and other appliances), hence the polarity only refers to energy system operation;
- The *provider/customer relationship* dimension (#8) ranges from being transaction-based in product-oriented S.PSSs, to relationship-based in result-oriented S.PSSs where a more intense relationship between provider and customers is created. For this reason, it can be aligned with the value proposition;
- Lastly, the *environmental sustainability potential dimension* (#9) can be encompassed in this group and it ranges from high (for result-oriented S.PSSs and use-oriented) to low (for product-oriented S.PSSs).

The horizontal axis encompasses the following dimensions:

- *Energy system* (dimension #1) and
- *Target customer* (dimension #7).

The *energy system* dimension (#1) focuses on the type of energy system, and includes stand-alone systems (mini-kit, individual energy system and charging station) and grid-based system (isolated mini-grid and connected mini-grid). For the purpose of this classification, the type of renewable source is not considered because this is transversal to the different types of energy systems. The *energy system* dimension is strictly related to *the target customer* dimension (#7). In fact: stand-alone systems, such as mini-kits and home systems, are targeted to individual users; S.PSSs offered through charging stations (e.g. lanterns sharing systems) are targeted to groups of users; finally, S.PSSs linked to mini-grids are offered to communities.

The resulting polarity diagram, combining the horizontal and the vertical axis, is visualised in Fig. 4.8. The vertical axis distinguishes six main types of S.PSS:

In product-oriented S.PSSs:

- *Pay-to-purchase with training, advice and consultancy services.* In this model, energy systems (with or without Energy-Using Products) are sold to the customer together with some advice related to the product/s sold, such as how to efficiently use the system, how to dispose of it, management training, etc. This advice can be delivered in many ways (e.g. after the purchase, during the use of the product, through training courses);
- *Pay-to-purchase with additional services.* Here, the provider sells the energy system but also offers services related to the installation, use and or end-of-life phases. These services can include a financing scheme, a maintenance contract, an upgrading contract, an end-of-life take-back agreement, etc.

In use-oriented S.PSSs:

- *Pay-to-lease.* In leasing models, the provider keeps the ownership of the system (and is often responsible for maintenance, repair and disposal), while the customer pays a regular fee for an unlimited and individual access to the leased product;

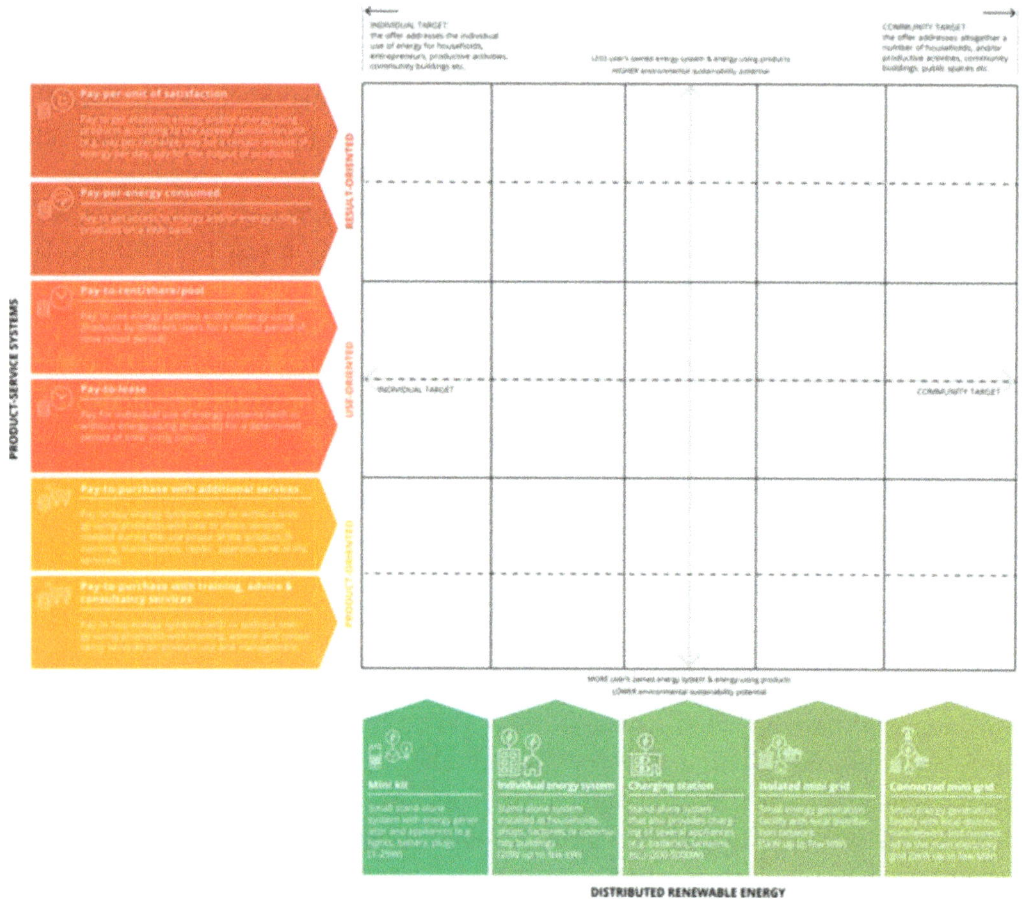

Fig. 4.8 Classification system. *Source* Emili [18]

- *Pay-to-rent/share/pool*. In this case, the provider keeps the ownership of the energy system and Energy-Using Products and is often responsible for maintenance, repair and disposal. Customers pay for the use of the Energy-Using Products (e.g. pay-per-hour) without having unlimited and individual access. Other clients in fact can use the product in other moments (different users can sequentially use the product).

In result-oriented S.PSSs:

- *Pay-per-energy consumed*. In this type of S.PSS, the provider offers a 'result' to customers and has the freedom of choosing the most appropriate technology to provide energy services. The energy solution provider keeps the ownership of the products (energy system and Energy-Using Products) and is responsible for maintenance, repair and disposal. Customers pay for the output of the system (energy) according to what they consume (pay-per-kWh);
- *Pay-per-unit of satisfaction*. Here, the provider offers access to energy (and/or Energy-Using Products) and customers pay according to the agreed satisfaction unit, e.g. pay-to-recharge products, pay for a certain amount of energy per day, pay for the output of products for a limited amount of time. The provider

chooses the best technology to provide the 'satisfaction' and keeps ownership and responsibility for the products (energy system and Energy-Using Products) involved.

Different from existing classification systems, this new system encompasses the majority of the dimensions characterising S.PSS and DRE, and thus provides an overview of the possible different models of S.PSSs applied to DRE. In other terms, it is a unified classification system capable of mapping and illustrating the different characteristics of these models.

It is important to highlight that the developed polarity diagram excludes some of the characterising dimensions: in particular the *capital financing* (#3) and the *organisational form* (#5) dimensions. Despite being crucially important for the implementation of S.PSS applied to DRE, they are cross-cutting to different types of offer models. In fact, the same type of offer model of S.PSS applied to DRE can be provided by different types of organisational forms and through different capital financing solutions. In other terms, these dimensions are not crucial for the classification system and for characterising offer models of S.PSS applied to DRE.

4.4.2 Archetypal Models of S.PSS Applied to DRE

After building the classification system, this was populated with 56 case studies. The aim of this activity was to understand the current situation in terms of existing S.PSS+DRE models. Cases were selected in order to cover, as much as possible, all the possible differences in the characterising dimensions (e.g. different types of technologies and energy sources, different types of target customers). The only common characteristic is the context of application: selected cases are related to rural areas in low- or middle-income contexts.

The next step was to group them into clusters of similar cases. This led to the identification of 15 archetypal models of S.PSS applied to DRE [17]. Cases included within each archetypal model present similar key traits, such as type of value proposition and target customer, but their secondary characteristics (e.g. the organisational form, the capital financing) are sometimes different. Figure 4.9 provides an overview of the 15 archetypal models.

The following text describes each archetypal model, coupled with a stakeholder system map, a visualisation showing the actors involved in the S.PSS offer and their relationships. For each archetype, a case study is illustrated. In the next paragraph, archetypal models are described starting from the bottom of the diagram: first product-oriented and then use-oriented and result-oriented S.PSSs (See Figs. 4.10, 4.11, 4.12, 4.13, 4.14, 4.15, 4.16, 4.17, 4.18, 4.19, 4.20, 4.21, 4.22, 4.23 and 4.24).

In product-oriented S.PSSs, the first group of archetypal models (1, 2 and 3) is related to pay-to-purchase with training, advice and consultancy services.

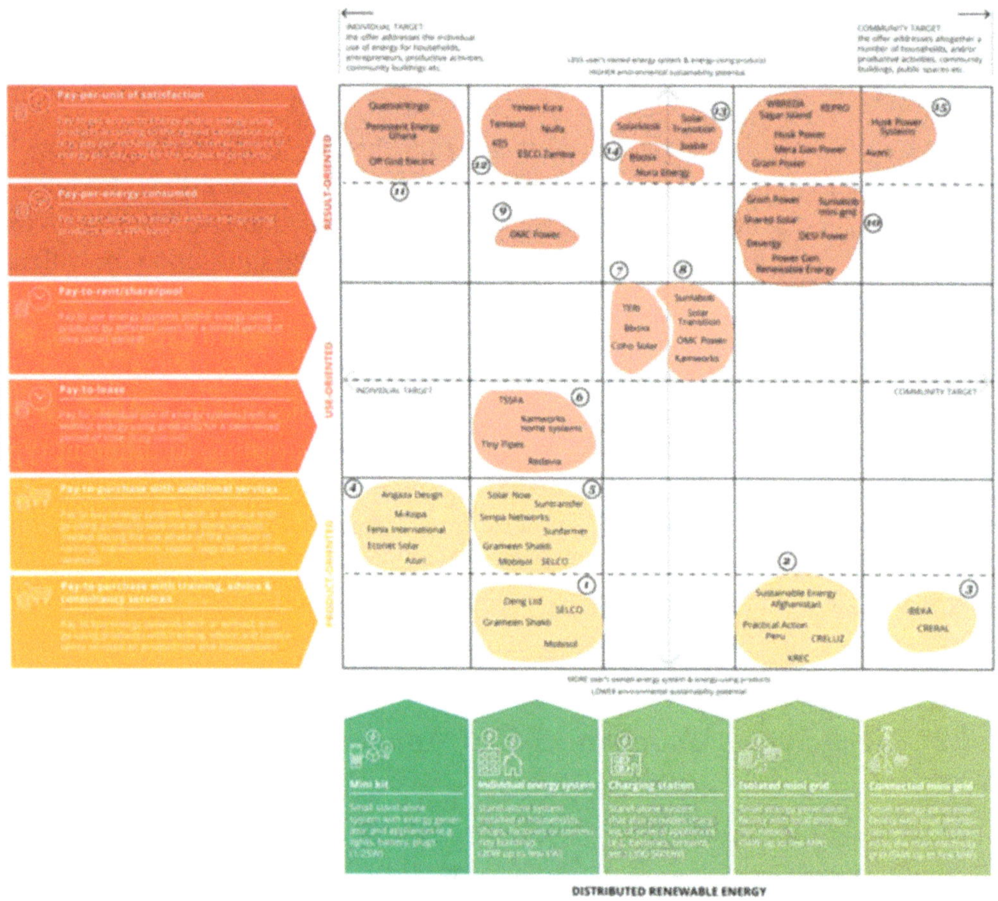

Fig. 4.9 Classification system with archetypal models. *Source* Emili [18]

1. *Selling individual energy systems with advice and training services.* In this model, the sale of individual energy systems (in most cases, solar home systems) is coupled with training and education. Depending on the target user, these services can focus on design, installation, repair and skills to develop a business on energy systems, or on basic maintenance and environmental awareness. Customers become owners of the systems at the moment of purchase and they are responsible for operation and maintenance.

Case study:

Mobisol/since 2010
Category: Solar Energy
Provider/s: Mobisol
Customers: Inhabitants
Location: East Africa

The company sells solar home systems with some additional services (financing, maintenance) and integrated mobile payment modality. Customers buy the chosen system and pay through mobile instalments over the credit period. The company

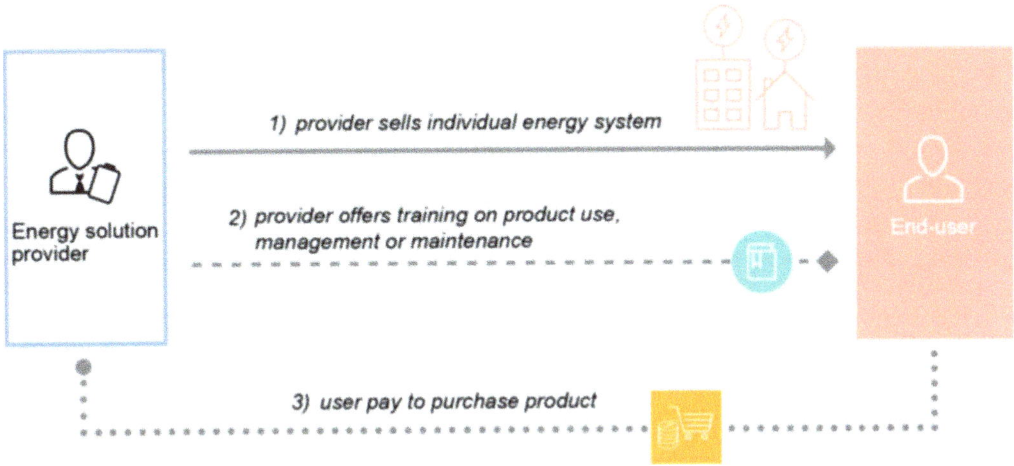

Fig. 4.10 Archetypal model 1: selling individual energy systems with advice and training services. *Source* Emili [18]

established the Mobisol Akademie, a training institution for staff, local entrepreneurs and contractors who wants to specialise in sales and technical support of solar home systems. The aim is to create local employment and capacity building and ensure that local expertise and assistance is provided.

2. *Offering advice and training services for community-owned and—managed isolated mini-grids.* The energy solution provider sells mini-grids to communities. Communities are responsible for operating and managing the system. They can also be in charge of designing a payment structure and fee collection. In addition to selling mini-grids, the provider offers a training service to a village committee on the operation, maintenance and management of the energy system. In some cases, communities may repay the installation with in-kind contributions such as labour.

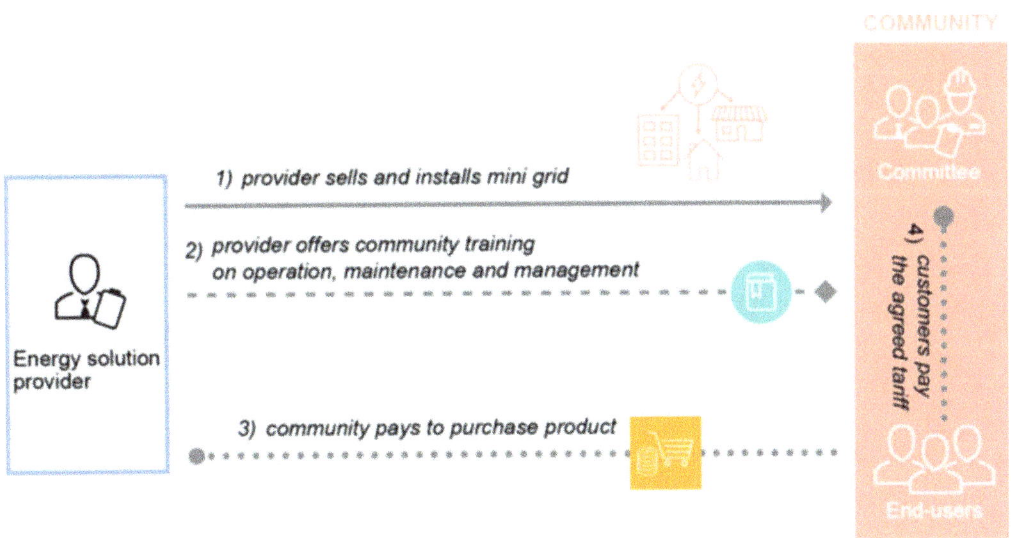

Fig. 4.11 Archetypal model 2: offering advice and training services for community-owned and community-managed isolated mini-grids. *Source* Emili [18]

Case Study:

Practical Action project
Category: Hydropower Energy
Provider/s: Practical Action, local manufacturers, village committee, local technicians
Customers: Local communities
Location: Perù

The Practical Action NGO helps communities in the Andes region in installing and setting up mini-grids running on hydropower. Practical Action partners with local manufacturers to design the system, then involves the communities by setting up a village committee that will take care of fee collection and trains some technicians who will perform daily operation and maintenance. The community participates in the system installation with construction labour and becomes owner of the energy system. End users pay for the electricity they consume with tariffs that differ between the types of customers.

3. *Offering advice and training services for community-owned and—managed connected mini-grids*. This model is very similar to the previous one but, in this case, the mini-grid is connected to the main electricity grid. In this case, the system allows the community to not only produce and distribute energy to the local network but also to sell electricity to the national electricity supplier.

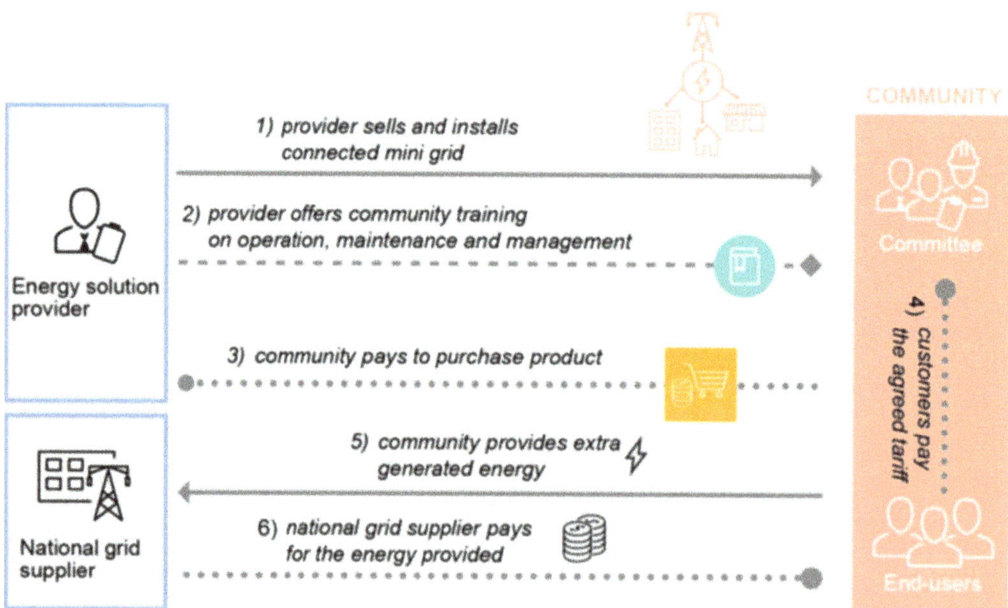

Fig. 4.12 Archetypal model 3: offering advice and training services for community-owned and community-managed connected mini-grids. *Source* Emili [18]

Case study:

Ibeka/since 2013
Category: Hydropower Energy
Provider/s: Ibeka, community-managed enterprise
Customers: Local communities
Location: Indonesia

IBEKA is a non-profit organisation that provides hydro mini-grids to communities with design, installation and community organisation. IBEKA sets up a community-managed enterprise to run the system and trains it for operation, maintenance and management. The grid-connected system allows communities to sell to the national grid supplier and revenues cover operation, maintenance, loan repayments and a community fund. End users pay according to the agreed tariff: pay-per-energy consumed (metre) or an agreed amount of energy per day.

The second group of product-oriented S.PSSs (models 4 and 5) is defined as pay-to-purchase with additional services.

4. *Selling mini-kits with additional services.* The provider sells mini-kits with additional services, such as financing, so that customers can pay through small, flexible instalments over time. After the credit period, usually 1 or 2 years, the ownership is transferred to the customer. Operation and maintenance are the customer's responsibilities and end users receive training on system care. During the credit period, the provider offers repair services and sometimes includes extended warranties after the credit repayment.

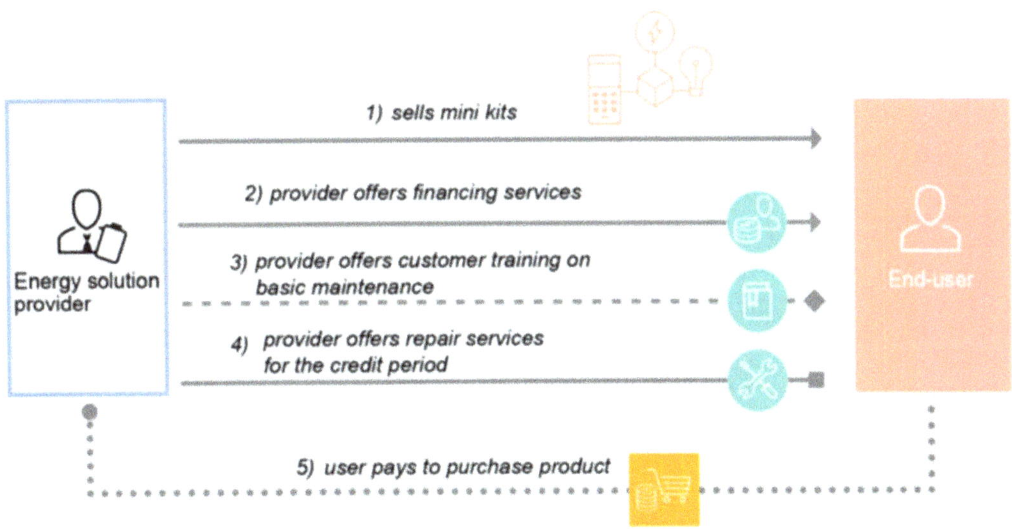

Fig. 4.13 Archetypal model 4: selling mini-kits with additional services. *Source* Emili [18]

Case study:

M-KOPA/since 2010
Category: Solar Energy
Provider/s: M-KOPA, d.Light, M-PESA
Customers: Inhabitants
Location: East Africa

M-KOPA, East Africa. M-Kopa provides energy by selling solar mini-kits with lights, radio, phone charging and enabling customers to pay small, flexible instalments over time. By partnering with a technology provider (d.Light) and using the existing network of mobile money M-PESA, the company allows customers to pay an initial deposit and then processes payments via mobile money transfer. If the payment does not occur, the system gets blocked. After the credit period, the customer owns the system and benefits from free and sustainable energy provision.

5. *Selling individual energy systems with additional services.* The provider sells individual energy systems with or without Energy-Using Products, and includes in their offer a range of services like financial credit, customer training, installation and aftersales services such as maintenance and repair. End users pay to purchase the energy system (with or without Energy-Using Products) and the ownership is transferred to them, sometimes after the credit period.

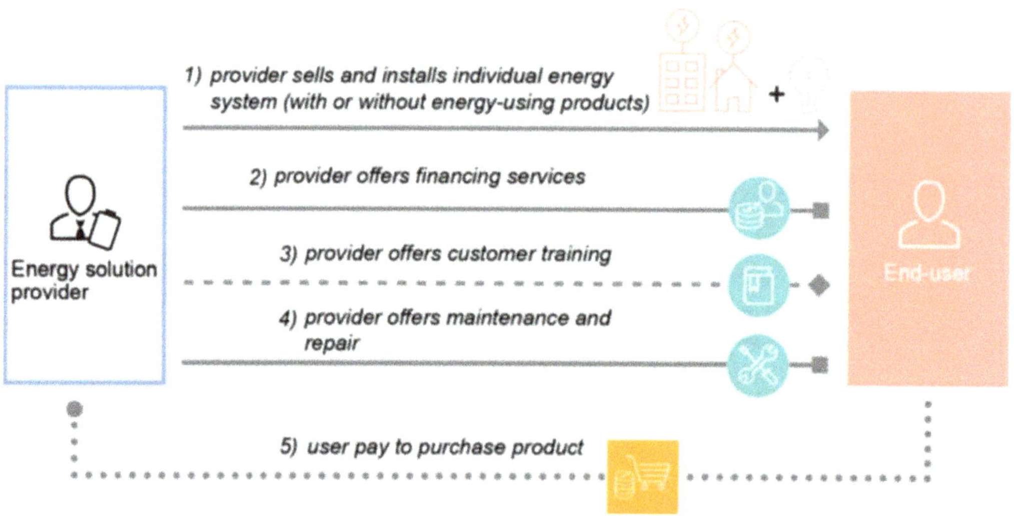

Fig. 4.14 Archetypal model 5: selling individual energy systems with additional services. *Source* Emili [18]

Case study:

Grameen Shakti/since 1996
Category: Solar Energy
Provider/s: Grameen Shakti, local technicians
Customers: Inhabitants
Location: Bangladesh

The company offers solar home systems with a service package inclusive of end-user credit, installation, maintenance and repair, take-back services. End users, low-income households and small businesses living in rural isolated communities, can purchase the product with microcredit services and be able to repay the loan in 3–4 years. To ensure an effective aftersale service, Grameen Shakti trains women as local technicians for repairs and maintenance of systems and for assemble solar accessories such as lamps, inverters and charge controllers.

Within the use-oriented S.PSSs group, we can distinguish between pay-to-lease (archetype 6) and pay-to-rent/share/pool models (archetypes 7 and 8).

6. *Offering individual energy systems (and Energy-Using Products) in leasing*. The provider offers energy home systems in leasing, with or without Energy-Using Products, for an agreed period of time. The offer may or not include Energy-Using Products. Customers do not become owners of the system but have unlimited access to it (and to the Energy-Using Products) during the leasing contract. Additional services, such as repairs and maintenance, are included in the product-service package.

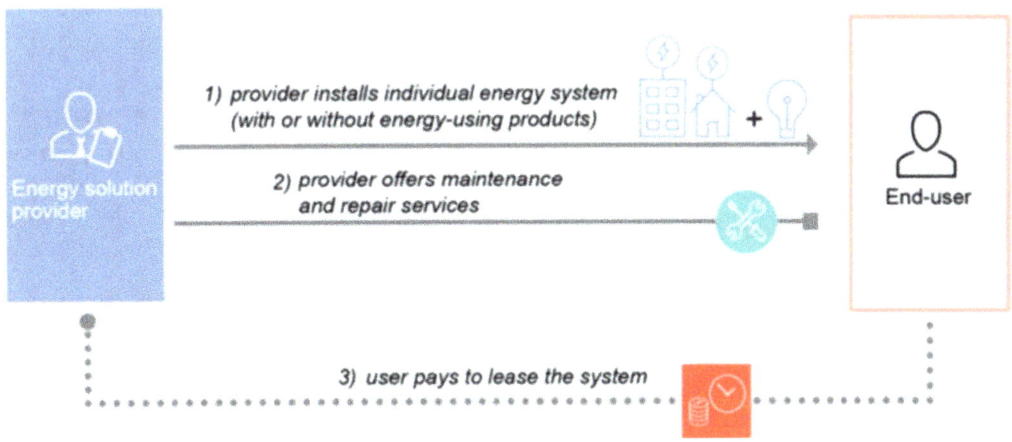

Fig. 4.15 Archetypal model 6: offering individual energy systems (and Energy-Using Products) in leasing. *Source* Emili [18]

Case study:

The Sun Shines for All/since 2001
Category: Solar Energy
Provider/s: The Sun Shines for All
Customers: Inhabitants
Location: Brazil

The company offers a solar home systems package (with Energy-Using Products) on leasing by providing customers with a contract that includes installation, maintenance, battery replacement after 3 years and take-back services. Users pay an initial deposit and a monthly leasing fee according to the system size and number of lights implied. The provider, who retains the ownership of systems and appliances, trains and employs local technicians who perform maintenance, repair and take-back services.

7. *Renting Energy-Using Products through entrepreneur-owned and—managed charging stations.* The charging station is sold to a local entrepreneur and ownership of both the charging station and the Energy-Using Products is transferred to him/her. Training on operation and management of the charging station is provided and financing services can sometimes be included. The local entrepreneur rents out the Energy-Using Products to end users, who pay a fee when they want to use the products involved. The entrepreneur is responsible for operation and maintenance of the system and the Energy-Using Products.

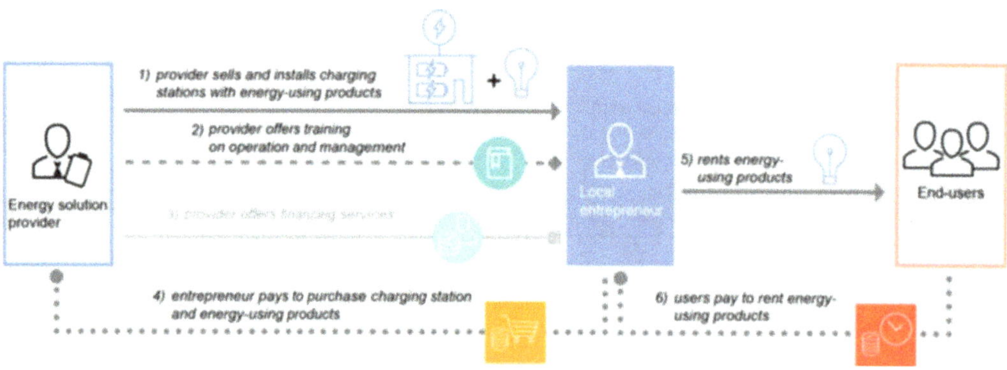

Fig. 4.16 Archetypal model 7: renting Energy-Using Products through entrepreneur-owned and entrepreneur-managed charging stations. *Source* Emili [18]

Case study:

Teri/since 2008
Category: Solar Energy
Provider/s: Teri, local entrepreneurs
Customers: Inhabitants
Location: India

TERI provides charging stations for renting lanterns to rural customers in India through an entrepreneur-led model. TERI sets up micro solar enterprises in un-electrified or poorly electrified villages. A local entrepreneur, who receives training and financing, buys and manages the charging station by renting the solar lamps every evening, for an affordable fee, to the rural populace. Every household pays a nominal charge (Rs. 2–4 approx.) per day per lantern for getting it charged.

8. *Renting Energy-Using Products through entrepreneur- or community-managed charging stations*. The energy solution provider instals a charging station for renting out Energy-Using Products to individual users. The provider keeps ownership of the charging system and the Energy-Using Products but the management and operation is undertaken by local entrepreneurs or by the community itself, who pays a leasing fee to use the charging station. End users pay to rent Energy-Using Products when they need.

Case study:

Sunlabob/since 2000
Category: Solar Energy
Provider/s: Sunlabob, local committee
Customers: Inhabitants
Location: Laos

The company provides energy services: it leases the charging station and Energy-Using Products (lanterns) to a village committee who in turns rents the products to the individual households. The committee oversees setting prices, collecting rents and performs basic maintenance. Sunlabob retains ownership,

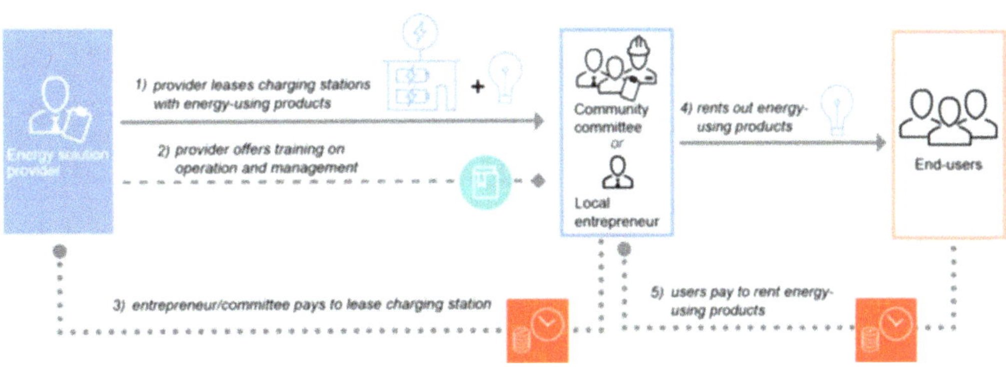

Fig. 4.17 Archetypal model 8: renting Energy-Using Products through entrepreneur or community-managed charging stations. *Source* Emili [18]

maintenance responsibilities and offers training services. End users can rent the recharged lantern for €0.2 and it will last for 15 h of light, while the committee pays to rent the charging station (€1.5 per month).

In result-oriented models, the first group of archetypal models (9 and 10) can be defined as pay-per-energy consumed.

9. *Offering access to energy (and Energy-Using Products) on a pay-per consumption basis through individual energy systems.* The provider instals individual energy systems at customers' site to satisfy the electricity need. Customers pay according to the energy they consume. The provider retains the ownerships of systems and takes care of operation, maintenance and repairs.

Case study:

Gram Power/since 2012
Category: Solar Energy
Provider/s: Gram Power, local entrepreneurs
Customers: Inhabitants
Location: India

Gram Power, India. Gram Power provides energy services in rural India through the installation and operation of mini-grids. Target customers are rural communities who get connected to the mini-grid and prepay for the energy they consume. Households get smart metres installed at their home and have the possibility to prepay electricity through local entrepreneurs. The entrepreneur, in fact, purchases in bulk energy credit from Gram Power, who keeps ownership of the system, and transfer the recharge into the consumer's smart metre through a wireless technology.

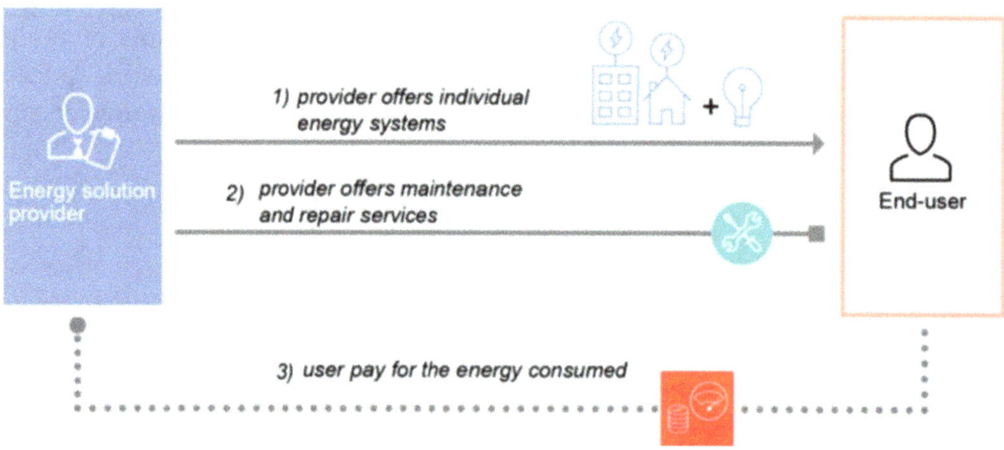

Fig. 4.18 Archetypal model 9: offering access to energy (and Energy-Using Products) on a pay-per-consumption basis through individual energy systems. *Source* Emili [18]

10. *Offering access to energy (and Energy-Using Products) on a pay-per consumption basis through isolated mini-grids.* The provider offers energy services by installing mini-grids (with or without Energy-Using Products) at a community level. End users pay according to the energy they consume. The provider always retains the ownership of the energy system and products involved. This model can present some variations (flows 5–8): in some cases, the local community or an entrepreneur receives training and can be involved in the management, operation and maintenance of the mini-grid or fee collection. In this case, end users pay their fees to the committee or entrepreneur, who is responsible for transferring them to the energy solution provider (in this case, flow 4 would then disappear).

Case study:

OMC Power/since 2000
Category: Hydropower/Wind/Solar Energy
Provider/s: OMC power
Customers: Telecommunication companies
Location: India

OMC Power offers energy solutions to productive activities (telecom tower companies) through large stand-alone power plants running on solar, hydro, wind or hybrid, according to the specific conditions. Mobile network operators get the power plant installed on site and pay according to the energy consumed (kWh). OMC Power retains the ownership of system and provides operation and maintenance.

The second group can be named pay-per-unit of satisfaction and encompasses archetypes 11–15.

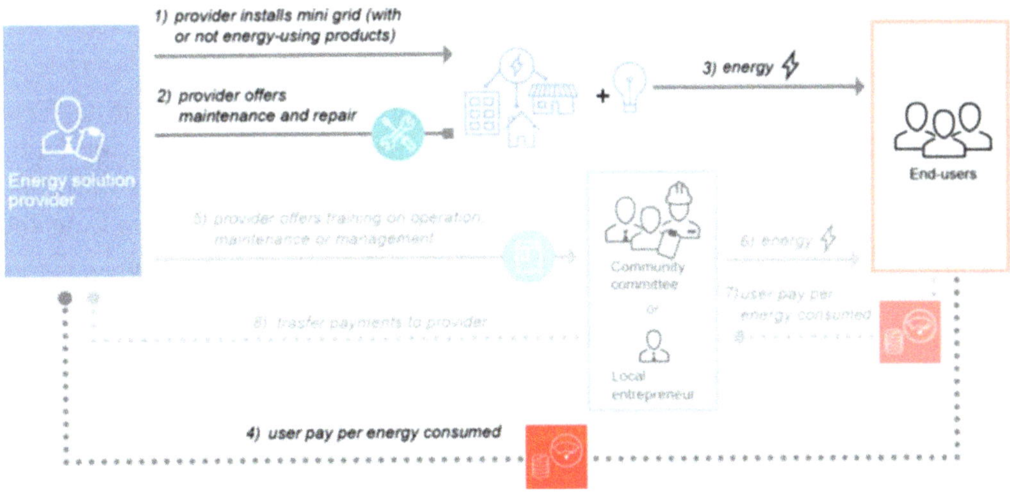

Fig. 4.19 Archetypal model 10: offering access to energy (and Energy-Using Products) on a pay-per-consumption basis through isolated mini-grids. *Source* Emili [18]

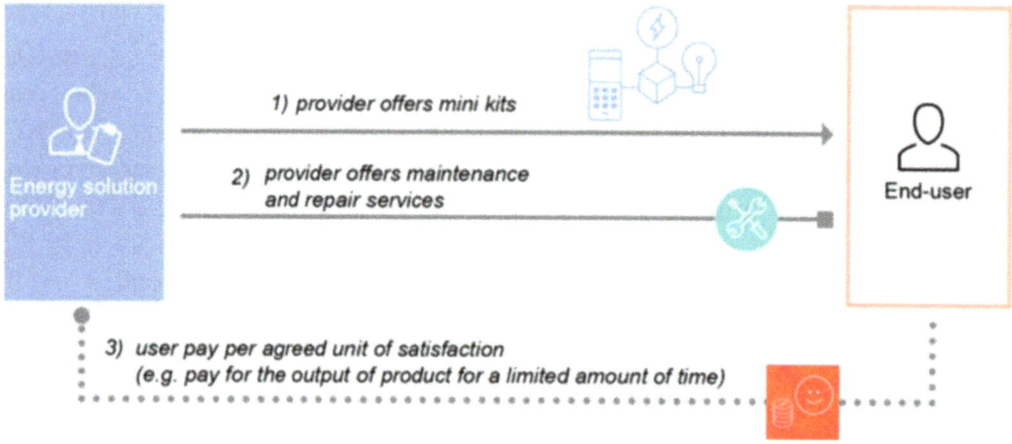

Fig. 4.20 Archetypal model 11: offering access to energy and Energy-Using Products on a pay-per-unit of satisfaction basis through mini-kits. *Source* Emili [18]

11. *Offering access to energy and Energy-Using Products on a pay-per unit of satisfaction basis through mini-kits.* The energy solution provider offers energy services through mini-kits equipped with Energy-Using Products. Users pay according to the service package they choose and the appliances they want to use (for example, they can pay to use two lights and a mobile charger for a maximum of 8 h a day). The provider, who retains ownership and responsibilities of the mini-kits, includes in the offer maintenance and repair services.

Case study:

Off-Grid Electric/since 2012
Category: Solar Energy
Provider/s: Off-Grid Electric, local entrepreneurs
Customers: Inhabitants
Location: Tanzania

Off-Grid Electric provides electricity services through solar mini-kits installed at customer's home. The service is tailored to users' needs and the satisfaction-based solution (two lights and a phone charger for tot hours/day) is paid by users with daily fees. Customers can choose the mini-kits with Energy-Using Products they want and upgrade with additional appliances. The starting kit includes two lights and a phone charger for 8 h a day. Off-Grid Electric retains ownership of systems and appliances and trains a network of local dealers for installation and customer support.

12. *Offering access to energy (and Energy-Using Products) on a pay-per unit of satisfaction basis through individual energy systems.* The provider instals energy home systems at the customer's site to provide electricity on a pay-per-unit of satisfaction basis. End users in fact pay a fixed monthly fee to get access to electricity or to use the included Energy-Using Products, usually for an agreed number of hours a day. The provider always retains the ownerships of the energy system (and Energy-Using Products) and takes care of maintenance and repairs.

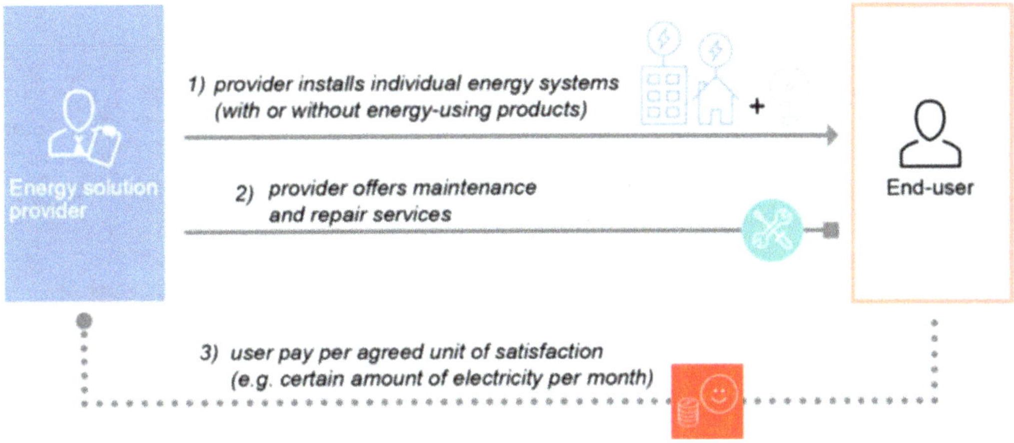

Fig. 4.21 Archetypal model 12: offering access to energy (and Energy-Using Products) on a pay-per-unit of satisfaction basis through individual energy systems. *Source* Emili [18]

Case study:

NuRa/since 2001
Category: Solar Energy
Provider/s: NuRa, local entrepreneurs
Customers: Inhabitants
Location: South Africa

NuRa, South Africa. NuRa provides energy through solar home systems. The company sets up an Energy Store where an entrepreneur is responsible for service provision and installation of the SHS. End users pay an initial fee (500R) and prepay a monthly fee of 61R that enables the connection of four fluorescent lamps and an outlet for a small black and white TV or a radio, operated on direct current (50 W panel) for four hours a day. Fees, based on the unit of satisfaction agreed (X amount of electricity for *X* hours a day), are collected through local businesses and shops. The ownership stays with NuRa, who is also in charge of maintenance and repairs.

13. *Offering access to Energy-Using Products through community- or entrepreneur-managed charging stations on a pay-per-unit of satisfaction basis.* The provider offers, together with training services, the charging station with Energy-Using Products to a local entrepreneur or a community committee. They in turn provide a range of energy-related services to end users, such as printing, purifying water and IT services to the local community. End users pay to get access to the Energy-Using Products (e.g. printer, photocopy or computer) on a pay-per-unit of satisfaction basis (e.g. pay-per-print or pay-per-unit of purified water). The entrepreneur/committee transfers part of the profits to the energy solution provider and is responsible for operation and maintenance of the charging station and Energy-Using Products.

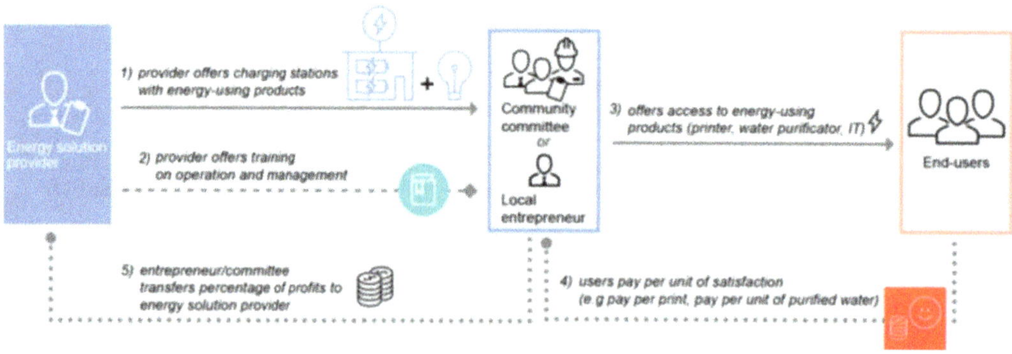

Fig. 4.22 Archetypal model 13: offering access to energy (and Energy-Using Products) on a pay-per-unit of satisfaction basis through individual energy systems. *Source* Emili [18]

Case study:

Solarkiosk/since 2011
Category: Solar Energy
Provider/s: Solarkiosk, local dealers
Customers: Inhabitants, local businesses
Location: Tanzania

As introduced (see paragraph 1.5) Solarkiosk targets local entrepreneurs, for the provision of energy services through charging stations. Due to the modular configuration of the station, Solarkiosk can provide a wide range of energy services such as Internet connectivity, water purification, copying, printing and scanning, etc. Customers pay for the agreed unit of satisfaction: pay-to-print, pay to get purified water, pay for Internet access, etc.

14. *Offering recharging services through entrepreneur-owned and entrepreneur-managed charging stations*. The technology provider sells, with training and sometimes with financing services, the charging station to a local entrepreneur who offers recharging services to customers. End users pay to recharge their products when they need (pay-per-unit of satisfaction), for example, they pay to charge mobile phones. The entrepreneur is owner of the system and responsible for operation and maintenance.

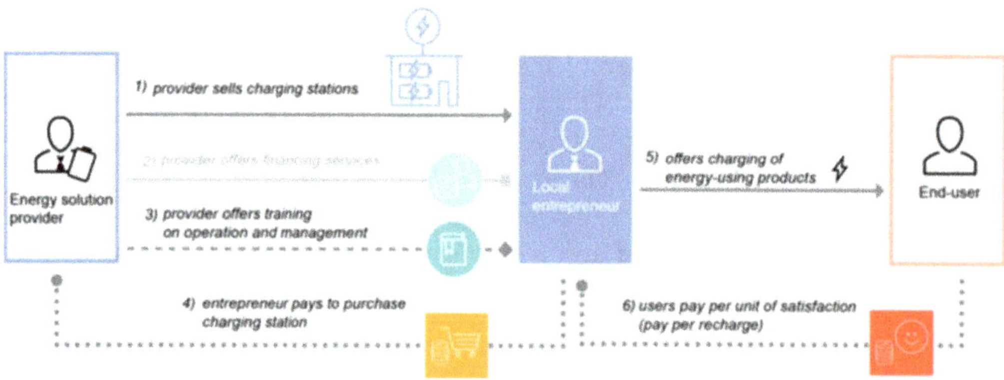

Fig. 4.23 Archetypal model 14: offering access to Energy-Using Products through community- or entrepreneur-managed charging stations on a pay-per-unit of satisfaction basis. *Source* Emili [18]

Case study:

Bboxx solar energy company/since 2010
Category: Solar Energy
Provider/s: Bboxx
Customer: Households
Location: Africa, Asia

Bboxx designs, manufactures, distributes and finances solar charging stations across Africa and Asia. One of their offers targets local entrepreneurs who buy the system (with credit services) and set up a phone charging business in their communities. Bboxx trains the entrepreneur in management and operation of the power station. End users pay per unit of satisfaction, in this case to get their phones charged.

15. *Offering access to energy (and Energy-Using Products) on a pay-per unit of satisfaction basis through mini-grids*. The provider offers energy services by installing mini-grids (and Energy-Using Products) at a community level. Mini-grids can be connected or not connected to the main grid. End users pay to get access to a limited amount of electricity for few hours a day. The provider always retains the ownership of the system and products involved in the offer. This model can present some variations (flows 5–9): in some cases, the local community or an entrepreneur is involved in the operation, management of the mini-grid, or in the fee collection as well. In this case, end users pay the agreed tariff to the community committee or entrepreneur and payments are then transferred to the energy solution provider (in this case, flow 4 would then disappear).

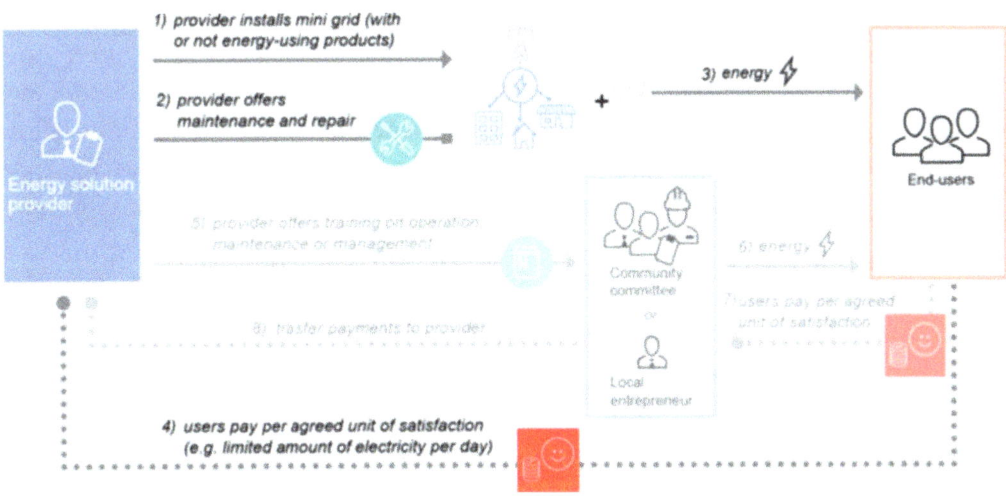

Fig. 4.24 Archetypal model 15: offering access to energy (and Energy-Using Products) on a pay-per-unit of satisfaction basis through mini-grids. *Source* Emili [18]

Case study:

Husk Power Systems (HPS)/since 2007
Category: Biomass Energy
Provider/s: Husk Power
Customer: Households and companies
Location: India

Husk Power Systems, India. The company provides energy solutions by designing and installing 25–100 kW isolated mini-grids based on biomass power plants. A partnership with local farmers is established to provide rice husk to power the plant. Households prepay a fixed monthly fee, ranging from 2 to 3€, to light up two fluorescent lamps and one mobile charging station. The company retains ownership and it employs local agents for operation, maintenance and fee collection.

In conclusion, this section described a new classification system for S.PSS applied to DRE which encompasses the seven major dimensions related to both S.PSS and DRE models. Through the empirical population of the classification system with 56 case studies, 15 archetypal models that describe the existing applications of S.PSS and DRE have been identified. It should be noted that the classification system can be easily updated adding new archetypal models. For this purpose, it is important to constantly integrate the latest state-of-practice in the classification system (i.e. collect new cases, position them in the map and identify new archetypes).

The classification system can be used by companies, practitioners and experts as strategic design tool. The different applications of the classification system are discussed in Sect. 6.2.

4.5 S.PSS Applied to DRE: Critical Factors

Scholars from various disciplines have been studying, over the past decades, how energy solutions in low- and middle-income contexts can be effectively and successfully implemented. A set of critical factors can be identified. The sections below provide an overview of the most important of these factors. In order to illustrate them in a clear and effective way, factors have been clustered in six main groups:

- **Customer**: it refers to the type of target customers addressed in the S.PSS solution;
- **Offer**: it refers to the different types of S.PSS+DRE models;
- **Products**: it refers to the *energy systems*, *renewable energy sources* and *Energy-Using Products* involved in the S.PSS solution;
- **Services**: it includes consultancy services (training, financing) and services provided during or at the end of the product life cycle (installation, maintenance and repair, product upgrade, end-of-life services);
- **Payment modality**: it refers to the different ways customers pay for the energy solution;
- **Network of providers**: it refers to the actors involved in providing the energy solution.

4.5.1 Customer

The design of S.PSS+DRE models must consider the complexity of the sociocultural context where these solutions will be implemented, as well as the customers' specific requirements, such as energy demand, awareness of technology, community organisation, customers' ability to pay etc. [61].

Energy demand and needs
When designing an energy solution, one of the first key factors to be defined is the energy demand of the customer, [64] in order to be then able to choose an appropriate technology to satisfy that demand [25]. Defining customers' demand means to identify the level of peak demand, how many hours electricity is used and the types of appliances run [64]. However, future increases of demand should also be predicted. To this regard, a common practice is to consider 30% extra capacity [48]. In addition, it is important to highlight that energy solutions must be customised not only considering the energy demand, but also the specific needs of the identified target customer [10, 50, 64].

Willingness to pay
It is also crucial to ensure that the solution is affordable and matches customers' willingness to pay. Willingness to pay is strongly related to customer awareness, expectations and perceived value of energy solutions [4]. For this reason, when offering an energy solution, it is important to enhance confidence in the technology through education and training on product use and benefits [4]. Adding perceived value to the energy solution is also important to improve willingness to pay. For example, adding extra appealing features, such as mobile charging sockets, incentivises users in setting up small income generation activities [26].

Ability to pay
Another critical factor to be considered is the ability to pay of low-income customers. A common practice is to adapt payment structures that mirror existing spending patterns of the target customer [4]. Offers that allow flexible payments according to seasonality of income and cash availability are an example of strategy to be adopted to enhance affordability [25]. In this context, it is also important to mention the role played by mobile payment technologies that can allow customers to pay small incremental amounts according to their income availability, mimicking existing spending patterns for non-renewable sources (kerosene, charcoal). Another common strategy is to partner with Microfinance Institutions (MFI) and provide financing services to end users and entrepreneurs [4].

More in general, affordability is also tackled by adopting use and result-oriented S.PSS models, where customers do not pay the full value of products but instead pay to get access to energy or Energy-Using Products.

Customer awareness and confidence
One of the barriers for introducing DRE technologies in BoP[2] contexts is related to the unfamiliarity or lack of awareness of renewable energy solutions. Therefore, it

[2]Bottom of the Pyramid (BoP) is the four billion people who live on less than say $3000 per year, or less than $2 a day.

is important to build confidence and trust in renewable energy systems and to communicate benefits of adopting these technologies. For example, marketing campaigns operating at different levels (word of mouth, radio, roadshows, partnering with existing brands) can help to achieve this goal [4]. The introduction of S. PSS models, and especially of ownerless solutions, can be problematic due to the cultural shift required in adopting new habits and behaviours contradicting the established norm of ownership [7, 22, 37, 41, 56]. It is therefore critical to educate customers on economic and environmental benefits derived from S.PSS innovations.

Recognise gender needs and address to equity
Designing energy services for BoP customers also requires understanding how energy impacts women and men differently and how their daily tasks, responsibilities and needs influence their electricity needs [39, 50]. Thus, it is important to address the different uses of energy for men and women and to favour the integration of women in the energy solutions. For example, this can be achieved by including women in the S.PSS management or in some roles such as technicians or entrepreneurs [1]. Enhancing their income generating skills through training activities is another strategy to be taken into consideration. In fact, women are more likely to afford energy services if it can be used to generate income, such as water pumping, husking and milling or home-based enterprises [1].

Involvement in the design and implementation process
Another key success factor for DRE projects is the involvement of users and communities as early as possible in the design as well as in the implementation process [10, 13, 23]. For example, the target customers can be involved in the design process by organising focus groups and adopting participatory approaches [23]. A good example of community involvement is provided by IBEKA, an Indonesian NGO that develops community-run mini-grid projects. The NGO works closely with the community in designing a tariff structure that covers operation and maintenance. Moreover, the NGO helps in setting up a community fund. This process ensures customers involvement and support, which is vital for the success of a project [13].

Differentiate the offer and address a mix of target customers
Ensuring financial sustainability can be problematic when targeting low-income customers. Thus, addressing a mix of customers including households, commercial and productive activities may be a recipe for success [10]. This can in fact ensure a more stable customer base. For example, a company could offer solutions to productive activities (which would represent its anchor customer), and at the same time deliver more affordable energy service to lower income users in the local community [38]. An example of this approach is provided by OMC Power, which supplies energy to telecom tower companies in rural India and instals charging stations to provide energy services to nearby communities.

4.5.2 Energy System

This section provides a summary of the key factors to be considered in relation to the physical elements of the energy system, including selection of renewable sources, DRE technologies and their applications.

Design for local conditions
Several factors need to be considered when the appropriate DRE technology, ranging from the environmental to the socioethical, economic, resource and regulatory aspects [31]. To begin with, the selection of a technology should reflect resource availability and be site specific [5, 35, 50, 64]. The technology should be also flexible and robust in terms of energy capacity and should consider energy demand changes and seasonality of resources [5, 6].

Selection of appropriate renewable sources
Each renewable energy source and respective technology has its own specific benefits, barriers and applications, which must be carefully considered when developing energy solutions. Main strengths and weaknesses of each type of renewable are discussed in Chap. 2.

Selection of appropriate energy configuration
Below, examples of renewable energy systems based on the presented structure and configurations (see paragraph 2.2) are provided.

Mini-kits
Mini-kits fit the configuration of distributed/stand-alone systems, and are small plug-and-play systems that include a small generator, lights, battery and other appliances such as radio or phone chargers [47]. Main strengths include easy installation, little maintenance required and low costs [47]. Because of its limited capacity, this technology is appropriate for households or small businesses, especially for scattered customers living in rural areas with a low-energy demand [13]. Usually these types of systems are coupled with mobile payments technologies, either providing microcredit or enabling pay-per-unit payments. Several examples are provided by companies such as M-Kopa, Azuri Technologies, Off-Grid Electric and Fenix International.

Individual energy systems
These are distributed/stand-alone systems that can be powered by solar, wind, hydro or biomass power and can target individual households, businesses or larger customers such as schools and productive activities. This type of technology suits especially off-grid customers and it is particularly convenient for the lack of transmission and distribution costs and for the flexibility to adapt to customers' needs [47]. However, individual systems require storage for extra-generated electricity and higher capital costs for customers (in the case the solution is not offered through use- or result-oriented S.PSS). Applications of this technology span from smaller solar home systems (e.g. Mobisol, Grameen Shakti, SELCO), to larger systems for productive activities (e.g. Redavia, OMC Power).

Charging stations
These are decentralised/stand-alone systems that can provide charging services (batteries, lanterns) and other services such as ICT or water purification. Main advantages are related to their mobility and flexibility, which makes them suitable for off-grid or emerging settings [47]. This technology has been usually applied in use-oriented S.PSS (pay-to-rent/share/pool) and in pay-per-unit of satisfaction models, enabling even lower income customers to have access to lanterns and batteries without paying upfront costs [9]. Larger charging stations can provide energy simultaneously to productive activities (as an individual energy system) and to nearby communities through the renting of appliances. This model has been implemented by an Indian company, OMC Power, which targets telecom tower companies in rural areas and villages nearby.

Isolated mini-grids
A mini-grid fits in the category of distributed/mini-grid and is a small generation facility that provides power through a local distribution network and it is not connected to the main grid [47]. This technology varies in applications, sizes and renewable sources used. Main advantages include: flexibility and adaptability to customers' demand; suitability for productive uses of energy and for multiple types of customers; enhancement to local development and employment as it can be managed and maintained by communities [35, 51]. Isolated mini-grids suits communities that are densely populated as they require enough demand for power to be profitable [50]. Main barriers for this technology are the need for skilled personnel for operation and maintenance, management and monitoring. In addition, they sometimes require specific regulatory frameworks and high capital financing [51]. Several players are providing energy through mini-grids, adopting different types of S.PSS models.

Connected mini-grids
Connected mini-grids fit in the category grid of mini-grids, and present further advantages compared to the isolated ones. First, they allow to sell electricity to the main grid; second, they can operate at higher load factors, thus enhancing economic sustainability [38]. This DRE system is particularly convenient for communities that live close to the national electricity grid or that may be connected in the near future, allowing the integration of the two energy supply systems [2]. S.PSSs involving connected mini-grids allow providers to have an anchor customer (national grid supplier) and distribute power to communities. Some examples can be found in community-owned and community-managed systems (e.g. IBEKA and CRERAL) and in pay-per-unit of satisfaction models (e.g. Avani and Husk Power Systems).

4.5.3 Services

PSS solutions usually include an articulated a set of services. These can range from training and consultancy services for product use and management, to financing and microcredit services, services that aim at extending the lifespan of products (installation, maintenance, repair, upgrade), and end-of-life services such as recycling or take-back.

Training services

A crucial factor to enhance the success of S.PSS+DRE solutions is to integrate the energy solution with training, consultancy and advice services [49]. These services can target different stakeholders: communities who will be responsible for managing the energy system; end users who need to learn how to properly use the product/s; local entrepreneurs and local technicians, who might be involved in providing maintenance and repair services.

Community training usually focuses on providing training in operation, maintenance and management of energy systems [30]. However, it must be highlighted that the delivery of these services should take into considerations the community's structure and its existing organisation [13]. To this end, it is suggested to discuss and agree on the provision of these services together with respected individuals and community leaders [13]. The involvement of local partners to provide training, such as NGOs or cooperatives that can deliver training in the local language, represents another potentially effective strategy [13, 23].

On the other hand, end-user training is crucial in order to ensure that customers understand capabilities and limitations of energy systems and optimise energy consumption (to reduce risk of blackouts and system failures). In fact, technical problems are often caused by systems' overuse, related to the lack of understanding of their limitations [10, 33]. This type of training services can be provided during system installation or through regular visits of technicians [33].

If the S.PSS+DRE solution also involve local entrepreneurs, it is important to empower them with training services. To this regard, coupling technical and business training with technologies that allow income generation can help fostering local economies and economic sustainability [45].

Establishing a network of local technicians who can provide prompt maintenance and repair services represents a fundamental aspect to ensure good after-contract services [23]. To this end, it is important to provide appropriate training to these technicians, focusing first on the most recurring technical challenges [23]. For example, Grameen Shakti (Bangladesh) trains women for performing repairs, maintenance and assembling of solar accessories, ensuring an effective after-contract service.

Microcredit to end users and entrepreneurs

Providing financial service to customers and entrepreneurs is an essential element to be integrated in solutions that targets BoP markets [32]. Microcredit services can be offered to customers with low or irregular income, and to local entrepreneurs who want to partner up with the energy provider. These services can be delivered in partnership with a Microfinance Institution (MFI) or other financial institutions. Crucial aspects to be considered when delivering these services are: willingness and ability to borrow; size of the down payment and monthly payments; and credit history and financing environment of target customers [16].

Installation

Providing installation as part of the S.PSS package is important in order to prevent that systems are installed improperly or wrong components are used [28]. Delivering installation services also provides an opportunity to train local technicians and end users [15].

Maintenance and repair

When providing a S.PSS solution, manufacturers have an economic interest to extend as much as possible the lifespan of the energy system and Energy-Using Products, in order to keep their costs as low as possible. For this reason, ensuring products long lifetime is essential to avoid system failures and improper repairs by end users, but also to reduce costs incurred by the provider. This is a crucial aspect, since it has been shown that the lack of a proper maintenance and repair network represents the main factor influencing the failure of community-managed systems. However, providing maintenance and repair can be challenging and expensive, in particular in rural and sparsely populated areas [4]. Common strategies include training local technicians in order to optimise service delivery, and using existing local infrastructures to store spare parts. For example, DESI Power (India), trains local entrepreneurs to operate and maintain power plants and adopts a standardised technology that does not require specialised skills.

Product upgrades

Product upgrading can be provided by offering modular and upgradable solutions, for example, allowing users to add elements over time (e.g. more lights, TV or radio). In this way, changes in consumers' wants and needs can be met modifying or upgrading the systems instead of manufacturing new products [42]. This is especially relevant for those S.PSSs in which the provider keeps the ownership and responsibility over the energy system and Energy-Using Products.

Also, replacing technologically obsolete components and products (e.g. batteries) can help optimising energy consumption. Again, this is aligned with the economic interest of providers who deliver result- and use-oriented S.PSSs.

Use optimisation services

The use optimisation of PSS can be provided as a service e.g. training on product/s use; or as technological solution, e.g. smart device to check SHS conditions. In the case of products which require energy in use, their use optimisation can reduce the use of resources (energy) and potentially toxic emissions. For example, Bboxx (Asia and Africa) provides SHS connected to its platform, which allows BBoxx to monitor energy consumption and the performance of the systems. These data are used to optimise products use and extend the life of the batteries. So forth, product use optimisation service can entail a gain sharing among the customer (reduced cost to reach her/his satisfaction due to low-energy use), the provider (reduced cost on energy and product replacement) and the environment (reduced use of resources and/or energy and materials).

End-of-life services
Providing services to ensure that the energy system and the Energy-Using Products are collected to be reused or remanufactured at the end of their lifespan is a key factor to ensure environmental sustainability [41]. In addition, as said before, when providing S.PSS solutions, manufacturers are economically incentivised in doing that, since they keep ownership of the equipment/products involved. End-of-life services can be provided through strategic partnerships with local actors which can collect broken equipment or expired batteries [23]. A key factor also relates to the design of the products involved in the S.PSS, which should be easily disassembled or designed to facilitate reuse and remanufacturing.

4.5.4 Network of Providers

PSS solutions usually involve a variety of different stakeholders in designing, producing and delivering the various element of the solution. The text below provides insights on the potential roles that different stakeholders can play in S. PSS&DRE solutions.

Private enterprise
Private enterprises can cover a variety of roles and be directly involved in the design, manufacturing and in the provision of services. Small-scale companies have the advantages in terms of proximity to customers, while larger scale enterprises may be more likely to ensure financial viability [33]. Independently from the size and structure of private companies, ensuring a strong local representation is considered a key success factor (Asian Development Bank; [23]). This can be achieved, for example, by involving agents that are part of the target communities and building a network of dealers and service personnel at a local context. In this context, existing networks related to other products (e.g. farms machinery, generators, telecommunications etc.) can be used to deliver the energy solution [24]. An example of this is the Kenyan company M-Kopa, which uses existing network of shops and retailers from its partner Safaricom to provide their energy solution.

Technology manufacturer
A key role in S.PSS solutions is covered by manufacturers. As already emphasised in Sect. 3.3 S.PSS Sustainability benefits, manufacturers should be part of the S. PSS solution in order to fully exploit the sustainability potentials offered by this model. In other words, manufacturers should keep ownership and/or responsibility over (some of) the life cycle stages of their products (energy system and/or Energy-Using Products). In fact, in these cases manufacturers have an economic interest in extending as much as possible the lifespan of their products, in order to reduce maintenance, repair and disposal costs, as well as the costs of manufacturing new products. For example, Kamworks, a Cambodian company, designs and manufactures solar home systems and lanterns and offers these products with a package of services: product-related services (maintenance, training) or advice and consultancy are offered when products are sold; in the use-oriented offer, energy

systems and Energy-Using Products are provided on renting or leasing, thus including all the required maintenance, repair and take-back services.

Community

The involvement of communities is a key factor in the success of energy solutions in BoP contexts [60]. However, it is important to highlight that they should be involved not only as consumers, but also as partners in the development and provision of the energy solution (ibid.). In fact, when directly involved in providing their own energy, communities have a strong incentive in operating and maintaining systems in a sustainable way [13]. The involvement should take place as soon and as much as possible from project implementation to the organisation of the energy solution [48]. To this end, a potential strategy is to involve established cooperatives or organisations at a village level in order to plan the energy solution according to the existing local organisational structure. However, community involvement could be hindered by their lack of technical and business skills. In these cases, communities, or their representatives, need to be properly trained, especially if they are involved in managing/delivering some aspects of the energy solution [13]. These training activities can be facilitated by partnering up with local NGOs.

Local entrepreneur

Local entrepreneurs are individuals, either with existing business activities or not, who can be involved in providing energy solutions or who can perform specific tasks such as maintenance services or fee collection. Local entrepreneurs might play an important role, especially when energy services have to be delivered in scarcely populated areas. For example, they can perform some services such as maintenance and repair or supporting product distribution. However, local entrepreneurs might usually need to be assisted with access to financing and microcredit [29, 50]. In fact, entrepreneurs may not be able to cover initial investments for setting up an energy business. In addition, it is important to consider that, depending on the activities the entrepreneurs have to perform, appropriate training should be provided [50].

Cooperative

Cooperatives are organisations composed by members that come together for a common purpose and can operate in various sectors (e.g. agriculture). They can provide energy solutions or play a role in partnership with the energy provider (similarly to what local entrepreneurs can do). The involvement of cooperatives is strategic because they have direct relationships with their members, they are characterised by self-regulatory forces and promote equal participation [63]. For example, successful cases are found in Nepal and Brasil, where cooperatives manage connected mini-grids and provide powers to local communities (e.g. CRERAL). Other important roles for cooperatives are: to provide financial support to end users and local entrepreneurs, as partners for the distribution of energy products or to support training, awareness campaigns [23].

Non-Governmental Organisation (NGO)

NGOs can be defined as mission-driven organisations that aim to achieve social or environmental objectives. The role of NGOs can be crucial in delivering S.PSS applied to DRE as they can be directly involved in providing (elements of a) solutions or can represent a strategic partner. For example, some NGOs such as Practical Action in Peru and Avani in India design and implement energy solutions at a community level and train villages on operation and maintenance of mini-grids. Their knowledge of the local context and their strong relationships with communities make NGOs a strategic partner in S.PSS+DRE solutions [24].

NGOs can also be involved in supporting some activities such as raising awareness, market research, or assisting in the distribution of products [3, 23]. In addition, through their network of donors and access to subsidies, NGOs can also facilitate customers' financing [38]. Also, they can help in selecting and training local entrepreneurs that will deliver energy solutions. For example Solar Sisters, an African NGO, partners with manufacturers such as d.Light and Angaza Design and empowers women by distributing solar technology through a network of franchisees. Solar Sister provides the women with a 'business in a bag', a start-up kit of inventory, training and marketing support.

Microfinance Institution (MFI)

Microfinance Institutions (MFI) are credit organisations that can play a key role in S.PSS+DRE as strategic partners for financing customers and entrepreneurs. For example, SELCO (Sri Lanka) offers tailored products and financing services to its clients by facilitating customers getting financed through its partners. SELCO partners with SEED (Sarvodaya Economic Enterprise Development Services) and while it focuses its expertise on providing high-quality services in installation and maintenance of systems, the MFI takes care of loans and repayments. When involving a MFI as financing partner, some key aspects must be taken into consideration. First, training and awareness must be provided to MFI staff in order to allow them to understand technology options and design credit offers accordingly [23]. Second, good communication and cooperation between MFIs and technology providers is essential in order to ensure fee collection and continuation of payments [34].

Public entities and governmental institutions

Other actors from the public sector (e.g. public utilities) might be involved in providing energy solutions or can be engaged as partners to cover some aspects of the S.PSS offer, such as financing or regulatory support. When large-scale utilities or ESCOs (Energy Service Companies) are responsible for providing energy services, they can cover all aspects from financing to marketing, to customer education and maintenance services [25]. This can be achieved thanks to their extensive experience, financial resources and technical capabilities [33, 48]. However, a key factor that should be considered in these cases is to ensure local presence and assistance to customers, for example, by training local entrepreneurs and technicians to provide maintenance and fee collection [33, 48].

In other cases, public entities can be partners for the project's financing, the provision of subsidies to customers or the creation of supporting policies. In fact, the regulatory aspects play an important role in facilitating or limiting the diffusion of S.PSS+DRE solutions, and governmental entities can contribute in creating appropriate protective policies and regulations [39].

4.5.5 *Offer*

As described in Sect. 4.4, S.PSS applied to DRE: a new classification system and 15 archetypal models, six types of S.PSS applied to DRE can be defined, and 15 Archetypal Models distinguish different types of S.PSS+DRE offers (see Sect. 4.4.2). In this section, main critical factors for each type of S.PSS offer are discussed.

Product-oriented: Pay-to-purchase
In this type of S.PSS and DRE, the ownership of energy system and appliances is transferred to the customer with additional services. This payment structure (pay-to-purchase) is usually adopted for small individual energy systems and mini-kits as investment costs are relatively low and the purchase includes the additional services provided [13]. In fact, access to financing is crucial when customers pay to purchase systems, because affordability is a critical aspect to be addressed [4, 10]. Companies offering microfinancing options (e.g. M-Kopa, Mobisol, Grameen Shakti, SELCo, Azuri Technologies, Fenix International etc.) have addressed this issues by spreading payments over a credit period.

When this model is applied to mini-grids, the S.PSS solution involves community-owned and community-managed systems and it usually implies in-kind contributions from the community or the involvement of subsidies/donations.

In terms of environmental sustainability, product-oriented S.PSSs present in general a lower potential compared to use and result-oriented offers [52, 53]. As highlighted before, S.PSS models must be properly designed to be a sustainable alternative to traditional business models. In this case, providers can offer advice on product use and training on energy system operation, aiming at optimising energy consumption. Additional services that aim at extending lifespan of products such as maintenance, and repair should be provided. Additionally, end-of-life services are crucial to ensure safe disposal of polluting and dangerous components, such as batteries [1, 50]. Companies such as Grameen Shakti and SELCO are succeeding in providing a complete service package, from installation to financing, maintenance and recycling of individual energy systems.

Use-oriented: Pay-per-time of use
In use-oriented S.PSSs two types can be distinguished: pay-to-lease and pay-to-rent/share/pool. With leasing, customers pay a regular fee (e.g. pay-per-month) for an individual and unique access to products. Leasing models should consider the ability to pay of customers, as users with unstable income may not be willing to sign for monthly payments if they are not sure they can afford it.

With renting, customers pay for the use of products for shorter periods of time (e.g. pay-per-hour, pay-per-day) and sometimes simultaneously with other users (pooling model). The ownership of equipment/products (and the responsibility for maintenance, repair, disposal etc.) is retained by providers. This model mimics the existing spending patterns of lower income customers (with kerosene) and customers pay only when they need or when they can afford the product [9].

These types of S.PSSs might trigger some rebound effects [8]. The impact on customers' behaviour should be considered when introducing ownerless solutions such as leasing and renting models. In fact, if the user does not own the product, he/she may adopt careless behaviours and misuse and mishandle of products can reduce their lifespan. Thus, products should be properly designed to be used and shared amongst different users [42]. In fact, since providers retain the products ownership, it is in their economic interest to have products that are long-lasting (easy to be maintained, repaired, upgraded), easy to remanufacture and easy to recycle.

Another aspect is related to the awareness of economic benefit emerging from adopting a use-oriented S.PSS. Customers may lack understanding about life cycle costs [59] and therefore prefer solutions where they become owners of the products.

Result-oriented: Pay-per-energy consumed
In result-oriented S.PSSs, consumption-based offers involve the provider retaining ownership of products (energy systems and Energy-Using Products) and the customer paying to get energy on a kWh basis.

Some issues related to this type of payment structure are related to the ability to pay of lower income customers and the process of fee collection [33, 46]. An effective approach is to monitor customers and conduct daily/weekly visits to ensure payments, such as DESI India and Gram Power solutions which involve local entrepreneurs working in villages and regularly visiting households. Another option is to use prepayment technologies to limit demand and avoid overconsumption [33]. However, when adopting this type of S.PSS, limited capacity of DRE systems must be considered. In fact, this type of offer may result in overconsumption and system failures, especially if customers are not aware of limits of renewable energy sources and technologies.

Some solutions allow extra capacity to be added according to energy demand. Shared Solar, for example, provides solar-based isolated mini-grids in Mali. When demand grows, additional PV panels are added to the generator and customers prepay for the energy they consume using mobile payments.

Result-oriented: Pay-per-unit of satisfaction
Another type of S.PSS applied to DRE has been defined as 'Pay-per-unit of satisfaction' and encompasses those models where customers pay to get access to energy and Energy-Using Products according to the agreed satisfaction unit. This S.PSS includes several payment structures:

- *pay-per-recharge*: pay a fixed cost for recharging an Energy-Using Product (e.g. a lantern or a phone).
- *pay-per-lux*: pay a fixed cost for an agreed level of luminance of a building.
- *pay-per-print or Internet connectivity*: pay a fixed amount to use Energy-Using Product/s.
- *pay-per-energy service package*: pay a fixed fee to have access to Energy-Using Products and a limited amount of energy.

In the case of pay-per-energy service package, customers might pay a fixed tariff according to the agreed result or according to the limits of the power generation [13]. In the first case, fees can be set according to different levels of consumption, determined on existing or desired appliances and the regularity of their use [27]. For example, customers pay to use few lights, a mobile charger and TV for a certain amount of time per day. In the second case, customers pay to have a limited agreed amount of energy per day. Here limiting devices such as smart metres can be used to ensure the energy provision is fixed and to avoid system overload. Some companies such as Off-Grid Electric in Tanzania are providing these unlocking systems to ensure payments are met. Others, such as Mera Gao Power in India, involve local entrepreneurs to collect weekly fees and ensure that the system automatically locks according to its generation capacity.

The inclusion of energy-efficient products is particularly crucial for this type of S.PSS and its application in low-income contexts [47]. Some studies highlighted that it is necessary to include energy-efficient components with the energy systems and ensure that users are lifted from the responsibility of replacing them [47]. The provider retains responsibilities for managing and operating on the products involved, avoiding that customers influence efficiency and capacity of the energy system. It has been argued that fixed tariffs do not encourage customers in conserving energy and avoiding overconsumption [2]. For this reason, it is crucial to ensure this aspect is tackled through technology (e.g. use of locking meters, inclusion of efficient Energy-Using Products) and through customer education and training. As discussed for use-oriented S.PSSs, solutions must be properly designed to deliver the sustainability potential and to provide a more environmentally friendly alternative to traditional business models.

Some barriers are also related to the applications of this type of S.PSS. The main cultural barrier is related to the adoption of ownerless solutions [22, 37, 41, 56]. In addition, in pay-per-unit/result models, the final user may feel less responsible for the good use of the system [33] and may tend to adopt careless behaviours [8]. In addition, as mentioned in use-oriented S.PSS, lacking understanding about life cycle costs may steer the choice towards solutions where they become owners of the products.

Mixed offers

Combining S.PSS offers can strategically mix payment structures for different customer segments. For example, lower income household can pay a fixed amount for a limited service (pay-per-unit of satisfaction) whereas productive activities or higher income customers can pay-per-energy consumed (kWh). This approach can ensure financial viability of S.PSSs [10] and provide customer's satisfaction according to specific needs of each target group. An example is provided by OMC Power, an Indian company that targets productive activities (telecom tower companies) on a pay-per-consumption basis and communities through a use-oriented model (renting of appliances).

4.5.6 Payment Channels

Different payment methods and channels can be adopted in S.PSS+DRE models. These include cash and credit, mobile payments, scratch cards and energy codes, in-kind contribution, fee collection and remote monitoring as an activity supporting payment.

Mobile payments
The wide adoption of mobile services and the great diffusion of mobile phones provide an opportunity to use this technology for payment purposes [12]. This type of payment tackles some of the main barriers of energy solutions at the BoP: revenue collection and affordability for customers [43]. In addition to representing represents an innovative way for low-income people to have access and pay for energy services, the integration of payments in mobile phones can also offer remote control of products' performances and consumption [43]. Several companies have adopted mobile solutions to collect (e.g. M-Kopa, Azuri Technologies, Mobisol and Shared Solar).

Scratch cards and energy credit codes
Scratch cards can be used to deploy energy credits in forms of unique codes that allow customers to prepay the electricity provision and unlock the energy system. This payment method can be very convenient for prepayment of systems and enables flexibility of payments, allowing users to mimic the patterns of airtime purchases [4, 23]. The involvement of local vendors and entrepreneurs in distributing prepaid cards is an important factor to be considered. Azuri Technologies, for example, sells mini-kits with a mobile credit service: after paying an installation fee, users purchase a scratch card at local vendors each week and adds credit to their unit via mobile phone.

Fee collection
The definition of a suitable fee collection scheme is extremely important as it can influence customers' willingness to pay [13]. To this end, an effective strategy is to adapt collection schemes to local income patterns (e.g. the seasonal income of farmers and rural customers) [33]. Ensuring local representation for the collection of payments, for example, involving local technicians who can perform regular visits to customers, represents another important factor [23]. For example, Mera Gao Power (Indian provider of mini-grids) involves local technicians who have existing relationships with customers and who visit households weekly to collect payments.

In-kind contribution
Another type of payment method can be in-kind contribution in the form of labour or resources produced by the end user or the community. Communities may be involved in providing labour for mini-grids works and construction [24]. This approach has also the added value of increasing the sense of ownership from the end users and is it critical to ensure a sustainable operation and maintenance [48]. Another example of including end users is to involve farmers who can provide biomass to generate power and have a reduction on their tariff [29].

Remote monitoring
Remote monitoring and metre reading are activities that support payment. Metres can be used to monitor energy consumption, disconnect non-paying customers and

load supply according to the contract agreement. Metres are also useful to incentivise energy consumption reduction and efficiency by allowing customers to have accurate record of their consumption.

The simplest option is to instal normal metres, but in these cases reading must be performed periodically (e.g. by technicians). Another option is to use smart metres and prepayment. In this way, the management of energy loads and payments can then be done remotely so that the problem of reading, billing and collecting can be solved [4, 43].

References

1. Adams S, Annecke W, Blaustein E, Jobert A, Proskurnya E, Nappez C et al (2006) A guide to monitoring and evaluation for energy projects. HEDON, 1–98
2. Africa Renewable Energy Access Program (AFREA) (2012) Institutional approaches to electrification—the experience of rural energy agencies in Sub-Saharan Africa. The World Bank Group, Washington DC
3. Bairiganjan S, Cheung R, Delia EA, Fuente D, Lall S, Singh S (2010) Power to the people. Investing in clean energy for the base of the pyramid in India. CDF, New Ventures, and WRI. Taramani
4. Bardouille P (2012) From gap to opportunity: business models for scaling up energy access. IFC World Bank Group, Washington DC
5. Barnett A (1990) The diffusion of energy technology in the rural areas of developing countries: a synthesis of recent experience. World Dev 18(4):539–553
6. Biswas WK, Bryce P, Diesendorf M (2001) Model for empowering rural poor through renewable energy technologies in Bangladesh. Environ Sci Policy 4:333–344
7. Ceschin F (2013) Critical factors for implementing and diffusing sustainable product-Service systems: insights from innovation studies and companies' experiences. J Clean Prod 45:74–88
8. Ceschin F (2014) Sustainable product-service systems. Between strategic design and transition studies. PoliMI Springer Briefs. Springer, UK
9. Chaurey A, Kandpal TC (2009) Solar lanterns for domestic lighting in India: viability of central charging station model. Energy Policy 37:4910–4918
10. Chaurey A, Krithika PR, Palit D, Rakesh S, Sovacool BK (2012) New partnerships and business models for facilitating energy access. Energy Policy 47:48–55
11. Correa HL, Ellram LM, Scavarda AJ, Cooper MC (2007) An operations management view of the service and goods mix. Int J Oper Prod Manage 27(5):444–463
12. Craine S, Mills E, Guay J (2014) Clean energy services for all: financing universal electrification. Sierra Club, San Francisco
13. Cu Tran Q (2013) ASEAN guidelines on off-grid rural electrification approaches. ASEAN Centre for Energy, Indonesia
14. Da Costa J, Diehl JC (2013) Product-service system design approach for the base of the pyramid markets: practical evidence from the energy sector in the Brazilian context. In: Micro perspectives for decentralized energy supply
15. Da Silva IP, Hogan E, Kalyango B, Kayiwa A, Ronoh G, Ouma AC (2015) "Innovative energy access for remote areas: the LUAV-light up a village" Project. In: Groh (ed) Decentralized solutions for developing economies—addressing energy poverty through innovation. Springer, UK
16. Energy Sector Management Assistance Program (ESMAP) (2001) Best practice manual. Promoting decentralized electrification investment. UNDP/The World Bank
17. Emili S, Ceschin F, Harrison D (2016) Product-Service Systems applied to Distributed Renewable Energy: a classification system and 15 archetypal models. Energy for Sustainable Development 32, 71–98

18. Emili S. (2017) Designing Product-Service Systems applied to Distributed Renewable Energy in low-income and developing contexts: A strategic design toolkit. PhD Thesis, Brunel University London
19. Friebe CA, Flotow P, Von Täube Fa (2013) Exploring the link between products and services in low-income markets—evidence from solar home systems. Energy Policy 52:760–769
20. Gaiardelli P, Resta B, Martinez V, Pinto R, Albores P (2014) A classification model for product service offerings. J Clean Prod 66:507–519
21. Gebauer H, Friedli T (2005) Behavioural implications of the transition process from products to services. J Bus Ind Mark 20(2):70–80
22. Goedkoop MJ, van Halen CJG, te Riele HRM, Rommens PJM (1999) Product service systems, ecological and economic basics. paper, Dutch ministries of Environment (VROM) and Economic Affairs (EZ)
23. Gradl C, Knobloch C (2011) Energize the BoP! Energy business model generator for low-income markets. A practitioners' guide. Endeva, Berlin
24. Gunaratne L (2002) Rural energy services best practices. USAID-SARI/Energy Program
25. Hankins M, Banks D (2004) Solar photovoltaics in Africa—experience with financing and delivery models. UNDP/GEF, New York
26. International Finance Corporation (IFC) (2011) The off-grid lighting market in Sub-Saharan Africa: market research and synthesis report. IFC/The World Bank, Washington DC
27. International Renewable Energy Agency (IRENA) (2012) Renewable energy technologies: cost analysis series. Solar photovoltaics. International Renewable Energy Agency, Abu Dhabi
28. International Solar Energy Society (ISES) (2001) Rural energy supply models. German Federal Ministry for the Environment, Nature Conservation and Nuclear Safety (BMU)
29. Iyer C, Sharma S, Khanna R, Laxman A (2010) Decentralized Distributed Generation for an inclusive and low carbon economy for India. India Infrastructure Report 2010, Infrastructure Development in a Low Carbon Economy
30. Jain S, Koch J (2009) Social entrepreneurship in the provision of clean energy: towards an organizing framework of market creation for underserved communities. Conference on Social Entrepreneurship, NY
31. Kishore VVN, Jagu D, Gopal EN (2013) Technology choices for off-grid electrification. In: Bhattacharyya (ed) Rural electrification through decentralised off-grid systems in developing countries, green energy and technology. Springer, London
32. Koch JL, Hammond AL (2013) Innovation dynamics, best practices and trends in the off-grid clean energy market. J Manage Glob Sustain 2:121–139
33. Lemaire X (2009) Fee-for-service companies for rural electrification with photovoltaic systems: the case of Zambia. Energy Sustain Dev 13:18–23
34. Lemaire X (2011) Off-grid electrification with solar home systems: the experience of a fee-for-service concession in South Africa. Energy Sustain Dev 15:277–283
35. Lemaire X (n.d.) Distributed generation: options and approaches. Sustainable Energy Regulation and Policy for Africa. REEEP and UNIDO
36. Mandelli S, Mereu R (2014) Distributed generation for access to electricity: "Off-main-grid" systems from home-based to microgrid. In: Colombo E et al (eds) Renewable energy for unleashing sustainable development. Springer, Heidelberg, pp 75–97
37. Manzini E, Vezzoli C, Clark G (2001) Product service systems: using an existing concept as a new approach to sustainability. J Des Res 1(2)
38. Martinot E, Chaurey A, Lew D, Moreira JR, Wamukonya N (2002) Renewable energy markets in developing countries. Ann Rev Energy Environ 27:309–348
39. Modi V, McDade S, Lallement D, Saghir J (2005) Energy services for the millennium development goals of the international bank for reconstruction and development. The World Bank, United Nations Development Programme, Washington DC
40. Mont O (2004) Product-service systems: Panacea or myth? PhD Dissertation. IIIEE, University of Lund, Sweden
41. Mont O (2002) Clarifying the concept of product—service system. J Clean Prod 10:237–245
42. Niinimäki K (2014) Sustainable consumer satisfaction in the context of clothing. In: Vezzoli et al (eds) Product-service system design for sustainability. Springer, Heidelberg

43. Nique M, Arab F (2012) Sustainable energy & water access through M2M connectivity. Report, GSMA Mobile for Development, London
44. Penttinen E, Palmer J (2007) Improving firm positioning through enhanced offerings and buyerseller relationships. Industrial Marketing Management, 36(5), 552–564
45. Practical Action (2016) Poor People's Energy Outlook—National Energy access planning from the bottom up. Practical Action Publishing Ltd. Rugby, UK
46. Roach M, Ward C (2011) Harnessing the full potential of mobile for off-grid energy. GSMA and IFC, London
47. Rolland S (2011) Rural electrification with renewable energy. Technologies, quality standards and business models. Alliance for Rural Electrification, Brussels, Belgium
48. Rolland S, Glania G (2011) Hybrid mini-grids for rural electrification: lesson learned. Alliance for Rural Electrification, Brussel, Belgium
49. Schillebeeckx SJD, Parikh P, Bansal R, George G (2012) An integrated framework for rural electrification: Adopting a user-centric approach to business model development. Energy Policy 48:687–697
50. Terrado E, Cabraal RA, Mukherjee I (2008) Designing sustainable off-grid rural electrification projects: principles and practices. The World Bank, Washington DC
51. The World Bank (2008) REToolkit: a resource for renewable energy development. Washington DC
52. Tukker A (2004) Eight types of product-service: eight ways to sustainability? Experiences from suspronet. Wiley Inter Sci 260:246–260
53. Tukker A, Tischner U (2006) Product-services as a research field: past, present and future. Reflections from a decade of research. J Clean Prod 14(17):1552–1556
54. Tukker A, van den Berg C, Tischner U (2006a) Product-services: a specific value proposition. In: Tukker A, Tischner U (eds) New business for Old Europe. Product services, sustainability and competitiveness. Greenleaf publishers, Sheffield
55. Tukker A, Tischner U, Verkuijl M (2006b) Product-services and sustainability. In: Tukker A, Tischner U (eds) New business for Old Europe. Product services, sustainability and competitiveness. Greenleaf publishers, Sheffield
56. United Nations Environmental Programme (UNEP) (2002) Product–service systems and sustainability. Opportunities for sustainable solutions. UNEP, Division of Technology Industry and Economics, Production and Consumption Branch, Paris
57. Vezzoli C, Ceschin F, Diehl JC (2015) The goal of sustainable energy for all. SV J Cleaner Prod 97:134–136
58. Vanitkoopalangkul K (2014) Sustainable Design Orienting Scenario for Sustainable Product-Service System (S.PSS) applied to Distributed Renewable Energy (DRE) in low and middle income (all) contexts, Design Master thesis, School of Design, Politecnico di Milano
59. White AL, Stoughton M, Feng L (1999) Servicizing: the quiet transition to extended product responsibility. Tellus Institute, Boston
60. Wilson E, MacGregor J, MacQueen D, Vermeulen S, Vorley B, Zarsky L (2009) Briefing. Business models for sustainable development. International Institute for Environment and Development (IIED), London
61. Wilson E, Wood RG, Garside B (2012) Sustainable energy for all? Linking poor communities to modern energy services. International Institute of Environment and Development
62. Wise R, Baumgartner P (1999) Go downstream: the new imperative in manufacturing. Harvard Bus Rev 77(5):133–141
63. Yadoon A, Cruickshank H (2010) The value of cooperatives in rural electrification. Energy Policy 3:2941–2947
64. Zerriffi H (2011) Rural electrification. Strategies for distributed generation. Springer Science +Business Media

Part II
System Design for Sustainable Energy for All

Chapter 5
Design for Sustainability: An Introduction

Historically, the reaction of humankind to environmental degradation, especially since the second half of the last century, has moved from an end-of-pipe approach to actions increasingly aimed at prevention. Essentially this has meant that actions and research focused exclusively on the de-pollution of systems have shifted towards research and innovation efforts aimed to reduce the cause of pollution at source.

In other words, the changes have been from: (a) intervention after process-caused damages (e.g. clean up a polluted lake), to (b) intervention in processes (e.g. use clean technologies to avoid polluting the lake), to (c) intervention in products and services (e.g. design product and services that do not necessitate processes that could pollute a lake), to (d) intervention in consumption patterns (e.g. understand which consumption patterns do not (or less) require products with processes that could pollute that lake).

Due to the characteristics of this progress, it is evident that the role of design in this context has expanded over time. This increasing role is due to the fact that: the emphasis shifts from end-of-pipe controls and remedial actions to prevention; the emphasis expands from isolated parts of the product life cycle (i.e. only production) to a holistic life cycle perspective; the emphasis passes further into the sociocultural dimension, into territory where the designer becomes a 'hinge' or link between the world of production and that of the user and the social/societal surroundings in which these processes take place; and the emphasis widens towards enabling users' alternative and more sustainable lifestyles.

Within this framework, the discipline of **Design for Sustainability** has emerged, which in its broadest and most inclusive meaning could be defined as:

a design practice, education and research that, in one way or another, contributes to sustainable development[1]

[1]Some authors adopt a more stringent definition of Design for Sustainability: e.g. Tischner [113] argues that Design for Sustainability requires generating solutions that are equally beneficial to the society and communities around us (especially unprivileged and disadvantaged populations), to the natural environment, and to economic systems (globally but especially locally).

C. Vezzoli et al., *Designing Sustainable Energy for All*,
Green Energy and Technology, https://doi.org/10.1007/978-3-319-70223-0_5

5.1 Evolution of Design for Sustainability

Design for Sustainability has enlarged its scope and field of action over time, as observed by various authors [23, 56, 93, 98]; Vezzoli and Manzini [20, 120]. The focus has expanded from the *selection of resources with low environmental impact* to the *Life Cycle Design* or *Eco-design* of products, to *designing for eco-efficient Product-Service Systems* and to *designing for social equity and cohesion*.

5.2 Product Life Cycle Design or Eco-Design

Since the 90s, attention has partially moved to the product level, i.e. to the design of products with low environmental impact. This attention was initially focused on redesigning individual qualities of individual products (e.g. reducing amount of material used in a product, facilitate disassembly, etc.). These early attempts to integrate environmental sustainability in product design go under the label of *green design* e.g. see [11]. It was only later, especially in the second half of the 90s, that this design approach broadened to systematically address the entire product life cycle, from the extraction of resources to the product end-of-life. This is usually referred as *product Life Cycle Design*, *Eco-design* or *product Design for Environmental Sustainability* [58]; [10, 75]; [112]; Hemel [44, 45]; ISO 14062 [50]; [99, 110]; Nes and Cramer [87]. In those years, the environmental effects attributable to the production, use and disposal of a product and how to assess them became clearer. New methods of assessing the environmental impact of products (the input and output between the technosphere, the geosphere and the biosphere) were developed; from among them, the most accepted is Life Cycle Assessment (LCA). In particular, two main approaches were introduced.

First, the concept of *life cycle approach*—from designing a product to designing the product life cycle stages, i.e. all the activities needed to produce the materials and then the product, to distribute it, to use it and finally to dispose of it—are considered in a holistic approach.

Second, the *functional approach* was reconceptualized from an environmental point of view, i.e. to design and evaluate a product's environmental sustainability, beginning from its function rather than from the physical embodiment of the product itself. It has been understood that environmental assessment, and therefore also design, must have as its reference the function provided by a given product. The design must thus consider the product less than the 'service/result' procured by the product.

In the late 90s design researchers also started to look at nature as a source of inspiration to address sustainability. One of these approaches is known as Cradle to Cradle (C2C) design [78], whose main principle 'waste equals food' focuses on creating open loops for 'biological nutrients' (i.e. organic materials) and closed loops for 'technical nutrients' (i.e. inorganic or synthetic materials). Different from product Life Cycle Design, C2C is mainly focused on the products' flow of material resources, and this might result in overlooking some other (and potentially more important) environmental aspects (e.g. energy consumption in the use phase).

As highlighted by Ceschin and Gaziulusoy [20], although product Life Cycle design focuses on the whole life cycle, this is mainly done from a technical perspective, with limited attention to the human-related aspects. Starting from the late 90s, design researchers started to address this issue by exploring design approaches that could complement product Life Cycle design. In particular, *emotionally durable design* [21, 22, 85, 117] focuses on the user-product emotional connection and proposes design strategies to strengthen that connection in order to extend product lifetime. On the other hand, *design for sustainable behaviour,* e.g. Lilley [64], [5], Lockton et al. [66] focus on the effects that users behaviour can have on the overall impact of a product, and on how design can influence users to adopt a desired sustainable behaviour and abandon an undesired unsustainable behaviour.

5.3 Design for Eco-Efficient Product-Service Systems

Even if it is true that the design approaches mentioned in the above section are fundamental to reduce the environmental impacts of products, from the end of the 90s we started to realise that a more stringent interpretation of sustainability requires radical changes in production and consumption models. For this reason, attention has partially moved to *design for eco-efficient Product-Service Systems*, a wider dimension than designing individual products alone [6, 9, 23, 25, 65, 74, 125]. From among several converging definitions, the one given by the United Nations Environment Programme [114] states that a Product-Service System (PSS) is 'the result of an innovative strategy that shifts the centre of business from the design and sale of (physical) products alone, to the offer of product and service systems that are together able to satisfy a particular demand'. In this context, it has therefore been argued [122] that the design conceptualization process needs to expand from a purely *functional approach* to a *satisfaction approach*, in order to emphasise and to be more coherent with the enlargement of the design scope from a single product to a wider system fulfilling a given demand related to needs and desires, i.e. a *unit of satisfaction.*[2]

Some design researchers have also proposed to adopt a territorial approach, looking at local socio-economic actors, assets and resources with the goal of creating synergistic linkages among natural and productive processes [2]. This approach has been labelled as *systemic design* [7, 8], and seeks to create not only

[2]This approach is further elaborated and declinated to the design of S.PSS applied to DRE as discussed in the first part of the book.

industrial products or S.PSSs but complex industrial systems, where material and energy flows are designed so that output from a socio-economic actor becomes input for another actor.

5.4 Design for Social Equity and Cohesion

Finally, design research has opened discussion on the possible role of *design for social equity and cohesion* [28, 76, 92], Mance [70], [13, 43, 73, 93, 121]; Carniatto and Chiara [14]; [33, 63]; Maase and Dorst [67]; [89]; Tischner and Verkuijl [111]; [27, 124]; dos Santos (2008); [122]. This potential role for design directly addresses various aspects of a 'just society with respect for fundamental rights and cultural diversity that creates equal opportunities and combats discrimination in all its forms' [35, 36]. Moreover, several writers and researchers urge a movement (and a key role for design) towards harmonising society such that it is not only just and fair but also that people are encouraged to be empathic, kind and compassionate for the benefit of others [38]; Rifkin (2010). We can indeed observe new, although sporadic, interest on the part of design research to move into this territory, to trace its boundaries and understand the possible implications.

Some researchers have adopted a bottom-up approach and investigated how people and communities innovate to address their own daily problems. 'Creative communities' [80] is an often used term to highlight the inventiveness of these ordinary people and communities (sometimes in collaboration with other local institutions, organisations and entrepreneurs) in designing, implementing and managing social innovations [53]. Typical examples include new forms of exchange and mutual help, community car-pooling systems, food networks linking consumers directly with producers, etc. Researchers in the field of *design for social innovation* have been exploring the characteristics of these innovations and the role of professional designers can play in supporting, promoting and scaling-up these community-based innovations, e.g. see [71].

Some authors have also focused on understanding how design can address social and environmental issues faced by people in low-income context, i.e. *design for the Base of the Pyramid* (*BoP*). The initial emphasis has been on product design for BoP, e.g. UNEP [26, 115]; dos Santos et al. (2009). More recently, the design research focus on BoP has moved to S.PSS, e.g. see [84]; Schafer et al. [102]; Jagtap and Larsson [51]; dos Santos [101], and social entrepreneurship and innovation, e.g. see [81]; Cipolla et al. [24].

Other authors [103, 122] have argued that a promising approach would be that of Sustainable Product-Service Systems (S.PSS) design for social equity and cohesion, or more shortly, System Design for Sustainability. This issue of Sustainable Product-Service System design for social equity and cohesion is described in the following chapter as in relation to the design of sustainable energy system accessible to all.

Nowadays, design for SE4A necessarily includes the issue of access to affordable, reliable, sustainable and modern energy for all, which UN has described in the

Sustainable Development Goals. In accordance with what was said before, design of S.PSS applied to DRE is called SD4SEA and it will be described in following sections.

5.5 Design for Socio-Technical Transitions

More recently, we understood that the challenge is not only to design sustainable solutions but also to identify which strategies and pathways are the most appropriate to favour and speed up their introduction and scaling-up [18, 20]. It has become in fact clear that some sustainable innovations (e.g. sustainable Product-Service Systems or sustainable social innovations) involve fundamental changes in culture, practice, institutional structures and economic structures, and thus they may cope with the current and dominant socio-technical systems [95]. For these reasons, a handful of design researchers have started to build upon system innovation and transition theories, e.g. see [41]; Kemp et al. [57, 94], to explore how design can address this issue. This resulted in an initial body of work exploring [20]: the development of a theory of design for system innovations and transitions [40]; how to design socio-technical experiments and transition paths [16, 18]; the connections between S.PSS design and system innovation theories [17, 19, 54, 55]; the importance of designing a multiplicity of interconnected and diverse experiments to generate changes in large and complex systems [52, 72, 79, 97]; the development of a curriculum on transition design for the first time Irwin et al. [49].

5.6 State of the Art of Design for Sustainability

Looking at the evolution of Design for Sustainability, it clearly emerges that there has been a widening in the scope of action. In particular, a number of considerations can be made [20].

First, DfS has broadened its theoretical and practical scope progressively expanding from single products to combinations of products and services to complex systems.

Second, this has been accompanied by an increased focus on the 'people-centred' aspects of sustainability. In fact, the first DfS approaches (e.g. see green design, eco-design, Cradle to Cradle) have predominantly focused on the technical aspects of sustainability. On the other hand, more recent approaches have recognised the crucial importance of the role of users (e.g. see emotionally durable design, design for sustainable behaviour), communities (e.g. see design for social innovation) and social dynamics in socio-technical systems (e.g. see design for system innovation and transition).

Third, a consideration can be made on the importance of each DfS approach. Even if it is true that sustainability must be addressed at a socio-technical system level, this does not mean that the approaches focusing at the product innovation level are less useful than systemic approaches. New socio-technical systems are anyhow characterised by a material dimension that needs to be appropriately designed using product innovation DfS approaches. Thus, each DfS approach is equally important because 'addressing sustainability challenges requires an integrated set of DfS approaches spanning various innovation levels, from products to socio-technical systems' [20].

5.7 Human-Centred and Universal Design

Introduction

This section discusses the importance of universal design and human-centred design approach in designing products, services, systems and environments. The aim is to design products, services, systems and environments that are usable, useful and desirable to a broad spectrum of people without the need for specialised designs for disabled users. The two approaches advocate for the concept of designing with diverse users with diverse characteristics rather than designing for users. That is, users are placed at the centre or core of all design activities. When universal design and human-centred design principles are applied, products, services, systems and environments meet the needs of potential users with a wide variety of characteristics. This can only happen when users are made active participants in the design process, and the possibility of their needs, interests and wants to be encoded in the final design are high and this may lead to the design to be accepted by many users without the need for adaptation or specialised design. The goal of universal design and human-centred design is to place a high value on diversity, equality, and inclusiveness of users when designing products, services, systems and environments.

Universal design

Universal design refers to a design approach that strives to ensure that products, services, systems and environments are usable by the broadest possible spectrum of people, without the need for adaptation or specialised design. When universal design principles are applied, products, services, systems and environments meet the needs of potential users with a wide variety of characteristics such as disabled or non-disabled, age, gender, capabilities or cultural background [12]. Universal design increases the potential for developing a better quality of life for a wide spectrum of users. Steinfeld and Maisel [106]; Petrie et al. [90] argue that it creates products, places and systems that reduce the need for special accommodation and many expensive hard to find assistive devices. The authors also advance that it reduces the stigma by putting users with disabilities on an equal playing field with non-disabled population. It also supports users in being self-reliant and socially engaged.

Universal design process

Burgstahler [12] proposed the following universal design process:

1. *Identify the application*—specify the product or environment to which you wish to apply universal design.
2. *Define the universe*—describe the overall population (e.g. users of service, product and system) and the diverse characteristics of the potential users of the design.
3. *Involve consumers*—involve users with diverse characteristics in all stages of the development, implementation and evaluation of the design.
4. *Adopt and apply guidelines or standards*—select existing universal design guidelines/standards and integrate them with other best practices in a given field.
5. *Plan for accommodation*—develop processes to address accommodation requests from users for whom the design does not automatically provide access.
6. *Train and support*—tailor and deliver constant training and support to stakeholders with respect to diversity, inclusion and practices that ensures accessibility, and inclusive of all users.
7. *Evaluate*—include universal design measures in periodic evaluations of the design, with a diverse group of users, and make modifications based on users' feedback.

Principles of universal design

According to the Centre of Universal Design [15], Ron Mace, Jim Mueller, Abir Mullick, Bettye Rose Connell, Mike Jones, Jon Sanford, Elaine Ostroff, Molly Story, Ed Steinfeld and Gregg Van der heiden collaborated to establish the principles of universal design to guide a wide range of design disciplines including environments, products and communications. This working group of architects, product designers, engineers and environmental design researchers proposed seven principles of universal design that can be applied to evaluate existing designs, guide the design process, and educate both designers and consumers about the characteristics of more usable, useful and desirable products, services and environments. The Centre of Universal Design [15] outlined the seven principles and guidelines of universal design, which are as follows:

1. **Equitable use**—the design is useful and marketable to people with diverse abilities.

   Guidelines:

   - *Provide the same means of use for all users: identical whenever possible; equivalent when not;*
   - *Avoid segregating or stigmatising any users;*
   - *Make the design appealing to all users.*

2. **Flexibility in use**—the design that accommodates a wide range of individual preferences and abilities.

Guidelines:

- *Provide choice in methods of use;*
- *Provide adaptability to the user's pace.*

3. **Simple and intuitive use**—use of the design is easy to understand, regardless of the user's experience, knowledge, language skills, or current concentration level.

   Guidelines:

   - *Eliminate unnecessary complexity;*
   - *Be consistent with user expectations and intuition;*
   - *Arrange information consistent with its importance;*
   - *Provide effective prompting and feedback during and after task completion.*

4. **Perceptible information**—the design communicates necessary information effectively to the user, regardless of ambient conditions or the user's sensory abilities.

   Guidelines:

   - *Use different modes (pictorial, verbal, tactile) for redundant presentation of essential information;*
   - *Provide adequate contrast between essential information and its surroundings;*
   - *Differentiate elements in ways that can be described (i.e., make it easy to give instructions or directions).*

5. **Tolerance for error**—the design minimises hazards and the adverse consequences of accidental or unintended actions.

   Guidelines:

   - *Provide fail safe features;*
   - *Discourage unconscious action in tasks that require vigilance.*

6. **Low physical effort**—the design can be used efficiently, comfortably and with minimum fatigue.

   Guidelines:

   - *Allow user to maintain a neutral body position;*
   - *Minimise sustained physical effort.*

7. **Size and space for approach and use**—appropriate size and space is provided for approach, reach, manipulation, and use, regardless of user's body size, posture, or mobility.

   Guidelines:

   - *Provide a clear line of sight to important elements for any seated or standing user;*
   - *Make reach to all components comfortable for any seated or standing user.*

The application of universal design in education is apparent in the following areas: Human-centred design, universal design for learning, universal design for instruction and universal design for education. In this chapter, the focus will be on human-centred design as it is more relevant to the overall objectives of Sustainable Energy for All by design.

Human-centred design

> When we dream alone, it is a dream. When we dream together, it is no longer a dream, but the beginning of reality [29].

This section discusses the concept of human-centred design as the process puts the user at the pinnacle of all design activities. This process is referred to as human-centred because it starts and end with the people one is designing for. The human-centred design process encourages the concept of designing with users rather than designing for users. The process commences by probing the needs, interests and behaviours of the users affected by the problem by listening and understanding their real needs. Human-centred approach contribute to innovation in design, increase productivity, improve quality, reduce errors, improve acceptance of new products and reduce development costs. This approach to design and development aims to make products, services and systems more useful, usable, pleasurable and cherisable. Some designers in new emerging economies have not yet embraced this approach in their practice, resulting in products, services or systems that do not respond to user's social, physical, emotional and cultural needs. Despite the advantages offered by this approach, it also has some limitations that need to be taken into consideration at the conceptual design stages. In this chapter, the authors opted to use the term human-centred design instead of user-centred design because the former suggests a concern for people, while the latter suggests a limited focus on people's roles as users.

Human-centred design is a methodology that puts users at the centre of the design process. It is an approach based on the needs and interests of users with special attention to making products, services or systems usable and understandable. Human-centred design is based on the premise that design is meaningful only when the focus of its activities and outcomes accommodate the largest possible number of people inclusive of their diversity [83]. It focuses on how people actually interact with specific products, services and systems, and designed environments, rather than prioritising the product form and appearance. IDEO [47] define human-centred design as a process and technique that create new solutions (products, services, systems, organisations, environments and modes of interaction) for the world. Human-centred design is an approach for designing products, services and systems, which are physically, perceptually, cognitively and emotionally intuitive [42]. Furthermore, the authors argues that the approach goes beyond the design's traditional focus on the physical, emotional and cognitive needs of users, and encompasses social and cultural factors. From these varied definitions, it is proposed that human-centred design is a multidisciplinary approach which is driven by users' needs and expectations, and at the same time involves users at every stage of the product development process in pursuit of creating useful, usable, engaging, pleasurable and desirable experiences. It has been noted that the above definitions emphasise the quality of the relationship between the person who uses the product to achieve some result and the product or service itself. The fundamental features of this relationship are effectiveness, efficiency, satisfaction and pleasure. The user-focused design concept, according to Stoll [107], has two characteristics: it satisfies people's needs in the most optimal way and it is superior to all competitive products, services and systems with respect to the design's characteristics.

The primary objectives of human-centred design, as argued by Rouse [96], are that: (a) the design should enhance a human ability, that is, user interests should be identified, understood and cultivated; (b) it should help overcome human limitations, for example, errors need to be identified and appropriate compensatory mechanisms devised and (c) it should foster user acceptance, that is, user preferences and concerns should be explicitly considered in the design process.

Figure 5.1. shows a human-centred design pyramid model proposed by Giacomin [42] which illustrates a journey from the more physical and physiological questions to the metaphysical questions. The model shows a hierarchy of human physical, perceptual, cognitive and emotional characteristics, followed gradually by more multifaceted, interactive and sociological considerations [42]. The model is made up of factors ranging from the physical nature of a user's interaction with the product, system and service to the metaphysical. The metaphysical meaning involves users forming their interpretation of the system, product and service based on its interaction with users. The metaphysical meaning is of paramount importance to social acceptance and commercial success. Giacomin [42] further argues that the designs whose characteristics answer questions which are high in the pyramid would be expected to offer a wider range of affordances to people and to embed themselves deeper within the user's culture.

Fig. 5.1 Human-centred design pyramid. *Source* Giacomin [42]

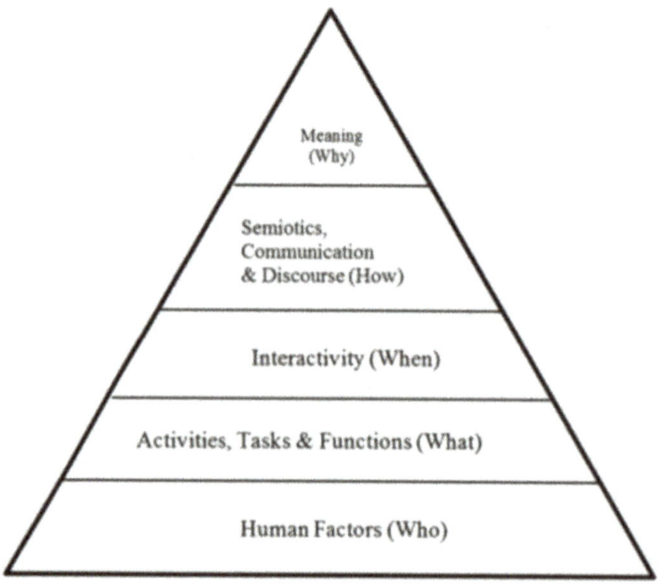

Human-centred design process

This study has adopted to use the human-centred design process in tackling complex energy challenges in new emerging economies. This approach was adopted because it can assist entities to connect better with the local people affected or dealing with energy issues. It can transform field data into actionable ideas, assist the team to find new opportunities and help to increase the speed and effectiveness of creating new solutions [47].

There are many models that represent the human-centred design process such as participatory design, ethnography, lead user approach, contextual design, co-design, co-creation and empathic design [62, 109, 119]; Beyer and Holzblatt [4]; Bennette [1, 3, 100]; EPICS [34, 47]; Steen et al. [105]. 'All the human-centred approaches have human beings in the process, involve users throughout the design process and seek to understand them holistically' [126]. Zoltowski et al. [127] state that it includes multidisciplinary collaboration to make products, services and systems useful, usable and desirable. In addition to the aforementioned, Krippendoff [61] identified the following features: (a) human-centred design employs both divergent and convergent thinking, (b) the process is concerned with how stakeholders attribute meaning through the use of the proposed design and (c) it includes the development of prototypes for the stakeholders to test their design ideas. According to the *International Organisation for Standardisation* 9241–210 [48], the human-centred design has six characteristics:

- The adoption of multidisciplinary skills and perspectives;
- Explicit understanding of users, tasks and environments;
- User-centred, evaluation-driven/refined design;
- Consideration of the whole user experience;
- Involvement of users throughout the design and development process and it is an iterative process.

This activity comprises the evaluation plan, data collection and analysis, reporting the results and making recommendations for change. One should iterate this activity until the usability and cherishability objectives are met.

One of the most widely used human-centred design models in tackling complex, wicked challenges, was developed by IDEO. IDEO's human-centred design process commences with a specific design challenge within a given context. The continuum of user involvement ranges from informative, through consultative and to participative Zoltowski et al. [127]. The process then goes through three main phases: Discover, Ideate, and Prototype.

i. **Discover**—*I have a challenge. How do I approach it? Who do I talk to?* This phase involves getting out into the world and learning from local people. The design team conducts field observational research by collecting stories and inspirations from the people.
ii. **Ideate**—*I learned something. How do I interpret it?* The design team conducts workshops and narrows down what has been learnt during the fieldwork, and translating those into themes, patterns and opportunities. During this phase, the

design team moves from concrete to more abstract thinking in identifying themes and opportunities, and then back to the concrete with solutions.

iii. **Prototype**—*I have an Idea. How do I build and refine it?* This phase involves rapidly evolving the design team's ideas into tangible designs based on real feedback. It also involves launching or implementing the proposed solution in the context it was designed to solve.

The participation of users is the main strength of human-centred design as they provide insight into the problem and this enhances the acceptance of the end product [69]. The approach requires that users should be actively involved throughout the design and development life cycle. Above all, this calls for designers to conduct immersive user research by watching users carrying out tasks in their own environment and asking open-ended questions about their actions, thoughts and feelings. This process is often accompanied by interviewing and video-recording users in their social context for later analysis and presentation to the design team.

The design team should be multidisciplinary, thus taking into account all knowledge and expertise required to produce a usable and pleasurable product or service. The cross-functional team might include all the relevant stakeholders who are directly or indirectly affected by the identified problem. The purpose of the team approach is to ensure that all needed information is readily available, as design decisions are made throughout the course of the project. Cross-functional teams are viewed as enhancing design creativity due to cross-fertilisation of thought processes, behaviour and functional skills. This team approach allows the development process to occur in non-linear iterations that bounce back and forth between disciplines, so that design decisions are fully informed. Such an approach provides a unifying framework and at the same time reduces the wastage of conflicting initiatives.

Consideration of sociocultural needs in human-centred design

Most designers tend to ignore the users' sociocultural needs when applying the human-centred design process in new emerging economies context. The evolution of design practice beyond ergonomics and human factors has been highlighted by Maguire [69], who argued for the need to identify stakeholders and contexts of use, and to apply creative processes. Gasson [39] highlights that 'user-centred system development methods fail to promote human interests because of a goal-directed focus on the closure of predetermined, technical problems'. The development of recognising the context and its people facilitated the probing, classification and description of the interactions, which occur between users and their environments and which has resulted in using personas and scenarios to provide a basis for describing users and contexts [86].

Users' culture is fundamental to the development of any new product or service as it plays a role in the acceptance of the product, service or system. Moalosi et al. [83] also argue that designs conceived from a sociocultural perspective may provide users with cultural meaning which facilitates their acceptance. Response to products often produces a mixture of intrinsic and extrinsic meaning. Products, services or systems are no longer seen only as functional objects, but they are seen for what they symbolise: their meaning, association and involvement in building a user's self-image. Therefore, the user's sociocultural needs should be considered in the early stages when the design is still relatively fluid and this provides a deeper insight and analysis of users' culture. It is envisaged that this type of design will lead to the creation of quality user cool experiences that add symbolic value to products, services or systems and to users' lives. This can also assist designers on how to create or design value, and to think of culture as a design resource. Krippendorff [60] sums it by saying, any design activity should identify the meaning which the product, service or system should offer to people.

Therefore, the field of human factors should extend beyond the usual physical and cognitive fit between products, services or systems and users, to embrace social and cultural considerations, personal needs, desires and aesthetic responses [104]. It is observed that human-centred design invites users to the design table, where they have traditionally been excluded. It seeks to bring the user closer to the designer, often reducing the step function of market research, which has tended to act as a barrier between the designer and the user [108]. In view of all these, the designing activity has been reshaped because it implies that ordinary people can contribute to the design process from the start. This methodology involves users in data gathering instead of relying on the designer's assumptions and experiences. The designer's perception has not been discredited, but only relocated to a more appropriate position. It can now be used to develop tools for understanding and facilitating creativity.

Human-centred design tools

The human-centred design toolbox techniques at times borrows from the fields such as psychology or sociology and sometimes those that emerge from design and engineering practice [30, 31, 42, 46, 116]. Human-centred design tools can be classified based on their intended use. The basic tools consist of facts about people such as anthropometric, biomechanical, cognitive, emotional, psychophysical, psychological and sociological data and models [42]. Such data often include materials on ergonomics or human factors which provide information about the abilities and limitations of users. Other tools consist of techniques for interacting with users to facilitate the discovery of meanings, desires and needs, either by verbal or non-verbal means. These techniques include ethnographic interviews, questionnaires, focus groups, participant observation and body language analysis. Table 5.1 summarises the human-centred design tools and the design phases that can be used.

Benefits and limitations of human-centred design

The benefits of usable and pleasurable products, services or systems include some of the following as identified by Wang [123] and Maguire [69]: Human-centred design (a) leads to increased productivity, that is, users concentrate on the task rather than the tool which could be causing a lot of problems, (b) reduces errors, (c) leads to reduced training and support, and yields products, services or systems that are easier to use and require less training, less user support (less documentation

Table 5.1 Human-centred design tools

Discover	Define	Ideation	Prototype	Testing
User observation, Cultural probes, Context mapping, Interviews, Questionnaires, Customer journey Mind mapping, Focus groups, Co-creation/co-design In-context immersion	Product collage, Personas Moodboards, Storyboards, Business model canvas Written scenario, Problem definition SWOT analysis	Analogies and metaphors, Synectics, SCAMPER, Bisociative technique, Brainstorming, Brainwriting, Who-what-where-when-why-how	Sketch modelling, Rapid prototyping, mock-ups, Empathy tools, Paper prototyping, Appearance models Quick-and-dirty prototypes, Experience prototyping	Scenario testing, User trials, Material testing, Safety testing

Source designed by the Authors

cost) and less maintenance, (d) enhances learning and user experience. Ultimately, all these lead to an improved acceptance through the trial and evolution of new products, services or systems before a full-scale launch. The approach enables an increased accessibility of products, services or systems to a range of users (for example, from an able-bodied to a disabled community).

In addition to the above benefits, human-centred design products, services or systems are viewed as having an improved quality, which makes them more competitive in a market that is demanding usable and pleasurable systems. Furthermore, other benefits include savings in developmental costs and time; increased trust in the product, service or system, as users are retained and new users are attracted; and increased job satisfaction for both the employer and employee, resulting in increased motivation and reduced stress. Human-centred design means relieving users of their frustration, confusion and a sense of helplessness [88], and helping them to feel in control and empowered. Moreover, IDEO [47] advance that some of the benefits of the human-centred design include: deep understanding of users' needs, development of customised solutions, facilitates bottom-up innovation, creates impact design, that is, desirable, feasible and viable, and user involvement is clearly useful and it has positive effects on both system success and in improving user satisfaction.

Despite the aforementioned benefits, human-centred design has limitations. Most scholars, for example, Rouse [96], Stanton [104] and Maguire [69], pay insufficient attention to the fact that this methodology has some restrictions. There is a problem in involving users in new innovative technologies: users of these technologies are not yet known, and therefore cannot be involved in the development process [59]. In support of previous point of view, Van Kleef et al. [118] and Marc [77] also argue that people may be unaware of their needs, unable to articulate their needs or unwilling to speak about their needs with an interviewer. In this case, innovative technologies refer to technologies that are either not yet realised at all or technologies that may be realised in a technical sense, but which are not part of the established social structure. Examples include interactive television and e-commerce software. These kind of products, services or systems are realised through the technology-centred approach, whereby the designer's expression of creativity is at the centre of the process.

The idea of user involvement is to engage people who are representatives of the assumed future users. However, if user requirements are fairly vague, it is difficult to determine who could be a representative of the future user. This creates a dilemma. If the scenario is still uncertain and it does matter which groups are going to be involved, the identity of the groups would remain uncertain. This condition is prone to outcomes which may not prove to be very reliable.

Potential users would not be willing to make an effort to participate in projects with uncertain outcomes and to cope with not yet fully determined technologies. Moreover, potential users rely on their previous work experience to contribute to the innovation process. If the new product is an invention, it becomes difficult for users to contribute fully because this is outside their experiences. This point of view is shared by Norman [88], who states that one cannot evaluate an innovation by asking potential users their views. This requires people to imagine something with which they have no experience. People find it difficult to articulate their real problems. Even if they are aware of the problem, they do not often think of it as a design issue. It is not possible to accurately predict user performance in future situations [91]. People do not react until the situation occurs; it is the context and environmental conditions that trigger their actions.

However, even if all the design problems are addressed, success is not guaranteed. In spite of this danger, even if the best-laid plans are suspect, by having put everything in place, the risk of failure has been reduced and there are better prospects of success [82]. In design, as in any other problem-solving process, it pays to analyse the problem before creating the solution. It is better to use 10% of the resources to find out how to use the remaining 90% properly than to use 100% of the resources the wrong way [37].

Summary

In this chapter, the importance of universal design and human-centred design with a bias towards the consideration of user sociocultural context have been emphasised, to enable designers to better understand and design for their intended users. Regardless of the research method used, the primary objective is to develop products, services, systems and environments for human diversity, social inclusion and equality. It also requires developing an understanding of users' values, attitudes and behaviour that can be translated into viable, powerful design concepts. In conclusion, universal design and human-centred design should not only include usability aspects but also it should go beyond and incorporate the cultural background and social situation of the user at the point of using the product, service, system or environment.

References

1. Atman CJ, Adams RS, Cardella ME, Turns J, Mosborg S, Saleem J (2007) Engineering design processes: a comparison of students and expert practitioners. J Eng Educ 96(4):359–379
2. Barbero S, Fassio F (2011) Energy and food production with a systemic approach. Environ Qual Manage 21(2):57–74
3. Bennett P (2006) Listening lessons: make consumers part of the design process by tuning in. Advertising Ages
4. Beyer H, Holtzblatt K (1998) Contextual design: defining customer-centered systems. Morgan Kaufmann, Burlington
5. Bhamra T, Lilley D, Tang T (2011) Design for sustainable behaviour: using products to change consumer behaviour. Design J 14(4):427–445
6. Bijma A, Stuts S, Silvester S (2001) Developing eco-efficient product-service combinations. In: Proceedings of the 6th international conference sustainable services and systems: transition towards sustainability? Amsterdam, The Netherlands, October 2001
7. Bistagnino L (2011) Systemic design: designing the productive and environmental sustainability. Slow Food Editore, Bra, Italy
8. Bistagnino L (2009) Design sistemico: Progettare la sostenibilita produttiva e ambientale. Slow Food Editore, Bra, Italy

9. Brezet JC, Bijma AS, Ehrenfeld J, Silvester S (2001) The design of eco-efficient services: method, tools and review of the case study based designing eco-efficient services. Dutch Ministries of Environment VROM, Delft University of Technology, Delft, The Netherlands
10. Brezet H, van Hemel C (1997) Ecodesign: a promising approach to sustainable production and consumption. Parigi, UNEP
11. Burall P (1991) Green design. Design Council, London
12. Burgstahler S (n.d.). Universal design: process, principles, and applications. Retrieved from http://www.washington.edu/doit/Brochures/Programs/ud.html
13. Carniatto V, Carneiro FV, Fernandes DMP (2006) Design for sustainability: a model for design intervention in a Brazilian reality of local sustainable development. In: Proceedings, International design conference, Dubrovnik, Croatia
14. Carniatto V, Chiara E (2006) Design for a fair economy. In: Proceedings of P&D conference, Curitiba, Brazil
15. Centre of Universal Design (1997) The principles of universal design. College of Design, NC State University. Retrieved from https://www.ncsu.edu/ncsu/design/cud/pubs_p/docs/poster.pdf
16. Ceschin F (2012) The introduction and scaling up of sustainable Product-Service Systems: a new role for strategic design for sustainability. Ph.D. thesis. Politecnico di Milano, Milan, Italy
17. Ceschin F (2013) Critical factors for implementing and diffusing sustainable Product-Service Systems: insights from innovation studies and companies experiences. J Clean Prod 45:74–88
18. Ceschin F (2014) Sustainable product-service systems: between strategic design and transition studies. Springer, London
19. Ceschin F (2014) How the design of socio-technical experiments can enable radical changes for sustainability. Int J Design 8(3):1–21
20. Ceschin F, Gaziulusoy IA (2016) Evolution of design for sustainability: from product design to design for system innovations and transitions. Des Stud 47:118–163
21. Chapman J (2005) Emotionally durable design. Objects, experiences, and empathy. Earthscan, London
22. Chapman J (2009) Design for (emotional) durability. Des Issues 25(4):29–35
23. Charter M, Tischner U (2001) Sustainable solutions: developing products and services for the future. Greenleaf Publishing, Sheffield, UK
24. Cipolla C, Melo P, Manzini E (2013) Collaborative services in informal settlements. A social innovation case in a pacified favela in Rio de Janeiro. In: NESTA ence, Glasgow Caledonian University London, 14–15 November 2013
25. Cooper T, Sian E (2000) Products to services, Friends of the earth. Centre for Sustainable Consumption, Sheffield Hallam University
26. Crul M, Diehl JC (2008) Design for sustainability (D4S): manual and tools for developing countries. In: Proceedings of the 7th annual ASEE global colloquium on engineering education, Cape Town, 19–23 October 2008
27. Crul M, Diehl JC (2006) Design for sustainability: a practical approach for developing economies. United Nations Environment Programme, Paris
28. Crul M (2003) Ecodesign in Central America. Delft University of Technology, Delft, the Netherlands
29. Denning S (2001) The Springboard: how storytelling ignites action in knowledge-era organisations. Butterworth-Heinemann, Oxford
30. Desmet PMA, Hekkert P (2007) Framework of product experience. Int J Design 1(1):57–66
31. Dunne A (2008) Hertzian tales: electronic products, aesthetic experience, and critical design. MIT Press, Cambridge, MA
32. Emili S, Ceschin F, Harrison D (2016) Product-service systems applied to distributed renewable energy: a classification system and 15 archetypal models. Energy Sustain Dev 32:71–98

33. EMUDE (2006) Emerging user demands for sustainable solutions final report. In: 6th Framework Programme (priority 3-NMP), European Community
34. EPICS Design Process (2009) Retrieved from https://sharepoint.ecn.purdue.edu/epics/teams/publicdocuments/epics_design_process.pdf
35. EU (2006) Renewed sustainable development strategy. Council of the European Union No. 10117/06, Brussels
36. EU (2009) Review of the european union strategy for sustainable development. Brussels
37. Friedman K (1997) Design science and design education. In: McGrory P (ed.), The challenge of complexity. University of Art and Design Helsinki Uiah, Helsinki
38. Fusakul SM, Siridej P (2010) DSEP: implementation of sufficiency economy philosophy in design' In Ceschin F, Vezzoli C, Zhang J (eds.) Sustainability in design: now! Challenges and opportunities for design research, education and practice in the XXI century. Proceedings of the Learning Network on Sustainability (LeNS) conference (vol 1). Bangalore, India, 29 September–1 October 2010. Greenleaf Publishing, Sheffield, UK
39. Gasson S (2003) Human-centered vs. user-centred approaches to information system design, J Info Technol Theory Appl (JITTA), 5(2):29–46
40. Gaziulusoy AI (2010) System innovation for sustainability: a scenario method and a workshop process for product development teams. Ph.D. Thesis, University of Auckland, Auckland, New Zealand
41. Geels FW (2005) Technological transitions and system innovations: a co-evolutionary and socio-technical analysis. Mass, Edward Elgar, Cheltenham, UK, Northampton
42. Giacomin J (2014) What is human centred design? Design J 17(4):606–623
43. Guadagnucci L, Gavelli F (2004) La crisi di crescita. Le prospettive del commercio equo e solidale. Feltrinelli, Milan
44. Heiskanen E (2002) The institutional logic of life cycle thinking. J Clean Prod 10(5)
45. Hemel CG van (2001) Design for environment in practice: three Dutch industrial approaches compared. In: 4th NTVA industrial ecology seminar and workshop: industrial ecology—methodology and practical challenges in industry
46. IDEO (2003) IDEO Method Cards: 51 Ways to Inspire Design. W. Stout Architectural Books, San Francisco, CA
47. IDEO (2011) IDEO human centered design toolkit for NGOs and social enterprises. Retrieved from http://www.ideo.com/work/item/human-centered-design-toolkit/
48. International Organization for Standardization (2010) ISO 9241–210: ergonomics of human-centred system interaction—Part 210: human-centred design for interactive systems. International Organization for Standardization, Geneva
49. Irwin T, Tonkinwise C, Kossoff G (2015) Transition design seminar Spring 2015. Carnegie Mellon Design School, Course Schedule
50. ISO 14062 (2002) Environmental management—integrating environmental aspects into product design and development. ISO/TR 14062:2002(E) ISO: Geneva
51. Jagtap S, Larsson A (2013) Design of product service systems at the base of the pyramid. In: Chakrabarti A, Prakash RV (eds), International conference on research into design, Chennai, India, ICoRD'13, Springer India
52. Jégou F (2011) Social innovations and regional acupuncture towards sustainability. Chinese J Design 214:56–61
53. Jégou F, Manzini E (eds) (2008) Collaborative services: social innovation and design for sustainability. Edizioni POLI.design, Milan
54. Joore P (2010) New to improve, the mutual influence between new products and societal change processes. Ph.D. thesis, Technical University of Delft, Delft
55. Joore P, Brezet H (2015) A multilevel design model—the mutual relationship between product-service system development and societal change processes. J Clean Prod 97:92–105
56. Karlsson R, Luttrop C (2006) EcoDesign: what is happening? An overview of the subject area of Eco Design. J Clean Prod 14(15–16)

57. Kemp R, Schot J, Hoogma R (1998) Regime shifts to sustainability through processes of niche formation: the approach of strategic niche management. Technol Anal Strateg Manag 10(2):175–195
58. Keoleian GA, Menerey D (1993) Life cycle design guidance manual: environmental requirements and the product system. EPA, USA
59. Korpela T (2002) Product modelling in early phases of the design process. In: Proceedings of the 7th international design conference, Dubrovnik, Croatia, May 13–17, 2002
60. Krippendorff K (2004) Intrinsic motivation and human-centred design. Theoretic Issues Ergonomics Sci 5(1):43–72
61. Krippendorff K (2006) The semantic turn: a new foundation for design. CRC/Taylor & Francis, Boca Raton
62. Leonard D, Rayport JF (1997) Spark innovation through empathic design. Harvard Bus Rev 75(6):102–113
63. Leong B (2006) Is a radical systemic shift toward sustainability possible in China? In: Proceedings, perspectives on radical changes to sustainable consumption and production (SCP), Sustainable Consumption Research Exchange (SCORE!) Network, Copenhagen
64. Lilley D (2007) Designing for behavioural change: reducing the social impacts of product use through design. Ph.D. thesis. Loughborough University
65. Lindhqvist T (2000) Extended producer responsibility in cleaner production. Doctoral dissertation, IIIEE Lund University, Lund
66. Lockton D, Harrison D, Stanton NA (2010) The design with intent method: a design tool for influencing user behaviour. Appl Ergonomics 41(3):382–392
67. Maase S, Dorst K (2006) Co-creation: a way to reach sustainable social innovation? In: Proceedings, perspectives on radical changes to sustainable consumption and production (SCP), Sustainable Consumption Research Exchange (SCORE!) Network, Copenhagen
68. Mace RL, Hardie GJ, Place JP (1991) Accessibility environments: towards universal design. Raleigh, New Carolina
69. Maguire MC (2001) Methods to support human centred design. Int J Hum Comput Stud 55 (4):587–634
70. Mance E (2001) A revolução das redes. A colaboracão solidária como una alternativa pós-capitalista à globalização atual. II ed., Editora Vozes: Petrópolis
71. Manzini E (2015) Design, when everybody designs. An introduction to design for social innovation. MIT Press, Cambridge, US
72. Manzini E, Rizzo F (2011) Small projects/large changes: participatory design as an open participated process. CoDesign 7(3–4):169–183
73. Manzini E, Jégou F (2003) Sustainable everyday: scenarios of urban life. Edizioni Ambiente, Milan
74. Manzini E, Vezzoli C (2001) Strategic design for sustainability. TSPD proceedings, Amsterdam
75. Manzini E, Vezzoli C (1998) Lo sviluppo di prodotti sostenibili. Rimini, Maggioli editore
76. Margolin V (2002) The politics of the artificial. University of Chicago Press, Chicago
77. Marc S (2011) Tensions in human-centred design. CoDesign 7(1):45–60
78. McDonough W, Braungart M (2002) Cradle to cradle: remaking the way we make things, 1st edn. North Point Press, New York
79. Meroni A (2008) Strategic design to take care of the territory Networking creative communities to link people and places in a scenario of sustainable development. Keynote presented at the conference "8° Congresso Brasileiro de Pesquisa e Desenvolvimento em Design", Campus Santo Amaro, San Paolo, Brazil
80. Meroni A (ed) (2007) Creative communities. People inventing sustainable ways of living. Edizioni Polidesign, Milan, Italy
81. Michelini L (2012) Social innovation and new business models: creating shared value in low-income Markets. Springer
82. Moalosi R (2000) Market research for student designers. DATA J Design Technol Edu 5 (3):239–244

83. Moalosi R, Popovic V, Hickling-Hudson A (2010) Culture-orientated product design. Int J Technol Des Educ 20(2):175–190
84. Moe HP, Boks C (2010) Product service systems and the base of the pyramid: a telecommunications perspective. In: Proceedings of the 2nd CIRP IPS2 Conference, 14–15 April 2010, Linköping, Sweden
85. Mugge R (2007) Product Attachment. Ph.D. Thesis. Delft University of Technology, The Netherlands
86. Mulder S, Yaar Z (2006) The user is always right: a practical guide to creating and using personas for the web. New Riders Publishers, Berkeley, CA
87. Nes van N, Cramer J (2006) Product life time optimization: a challenging strategy towards more sustainable consumption patterns. J Clean Prod 14(15–16)
88. Norman DA (2004) Emotional design: why we love (or hate) everyday things. Basic Books, New York
89. Penin L (2006) Strategic design for social sustainability in emerging contexts. Ph.D. thesis, Politecnico di Milano, Milan
90. Petrie H, Darzentas J, Walsh T (2016) Universal design 2016: learning from the past, designing for the future. IOS Press, Amsterdam
91. Popovic V (2002) Activity and designing pleasurable interaction with everyday artifacts. In: Pleasure with products: beyond usability in Jordan PW, Green WS (eds), Taylor and Francis, London, pp 367–376
92. Razeto L (2002) Las empresas alternativas. Nordam: Montevideo, Uruguay
93. Rocchi S (2005) Enhancing sustainable innovation by design: an approach to the co-creation of economic, social and environmental value, doctoral dissertation. Erasmus University, Rotterdam
94. Rotmans J, Kemp R, Van Asselt M (2001) More evolution than revolution: transition management in public policy. Foresight 3(1):15–31
95. Rotmans J, Loorbach D (2010). Towards a better understanding of transitions and their governance: a systemic and reflexive approach. In: Grin J, Rotmans J, Schot J (eds) Transitions to sustainable development. new directions in the study of long term transformative change. Routledge, London
96. Rouse WB (1991) Design for success: a human-centered approach to designing successful products and systems. Wiley, New York
97. Ryan C (2013) Eco-acupuncture: designing and facilitating pathways for urban transformation, for a resilient low-carbon future. J Clean Prod 50:189–199
98. Ryan C (2004) Digital eco-sense: sustainability and ICT—a New Terrain for innovation. Lab 3000, Melbourne
99. Ryan C (2003) Learning from a decade (or So) of eco-design experience (part one). J Ind Ecol 7(2)
100. Sanders EBN, Stappers PJ (2008) Co-creation and the new landscapes of design. CoDesign 4(1):5–18
101. Santos A, Sampaio CP, Giacomini da Silva JS, Costa J (2014). Assessing the use of Product-Service Systems as a strategy to foster sustainability in an emerging context. Prod Manage Develop 12(2):99–113. doi:http://dx.doi.org/10.4322/pmd.2014.012
102. Schafer C, Parks R, Rai R (2011) Design for emerging bottom of the pyramid markets: a Product Service System (PSS) based approach. In: Proceedings of the ASME 2011 international design engineering technical conferences & computers and information in engineering conference IDETC/CIE 2011, 28–31August 2011, Washington, DC, USA
103. Soumitri V, Vezzoli C (2002) Product service system design: sustainable opportunities for all. A design research working hypothesis. Clothing care System for Kumaon Hostel at IIT Delhi, In: Proceedings, Ecodesign international conference, New Delhi
104. Stanton N (1998) Product design with people. In: Mind in Stanton N (ed), Human factors in consumer products, Taylor and Francis, London, pp 1–17
105. Steen M, Manschot M, De Koning N (2011) Benefits of co-design in service design projects. Int J Design 5(2):53–60

106. Steinfeld E, Maisel J (2012) Universal design creating inclusive environments. Wiley, New Jersey
107. Stoll HW (1999) Product design methods and practices. Marcel Dekker Inc, New York
108. Storer I, McDonagh D (2002) Embracing user-centred design: the real experience. In: McCabe PT (ed) Contemporary ergonomics. Taylor and Francis, London, pp 309–313
109. Suchman LA (1987) Plans and situated actions: the problem of human–machine communication. Cambridge University Press, Cambridge
110. Sun J, Han B, Ekwaro-Osire S, Zhang HC (2003) Design for environment: methodologies, tools, and implementation. J Integrated Design Process Sci 7(1)
111. Tischner U, Verkuijl M (2006) Design for (social) sustainability and radical change, In: Proceedings, perspectives on radical changes to sustainable consumption and production (SCP), Sustainable Consumption Research Exchange (SCORE!) Network, Copenhagen
112. Tischner U, Schmincke E, Rubik F (2000) Was ist EcoDesign. Birkhauser Verlag, Basel
113. Tischner, U. (2010) 'Design for Sustainability: where are we and where do we need to go?' In Ceschin, F., Vezzoli, C. and Zhang, J. (eds) Sustainability in design: now! Challenges and opportunities for design research, education and practice in the XXI century. Proceedings of the Learning Network on Sustainability (LeNS) conference (vol. 1). Bangalore, India, 29 September-1 October 2010 (Sheffield, UK: Greenleaf Publishing)
114. United Nations Environmental Programme, UNEP (2002) Product-Service Systems and Sustainability. Opportunities for Sustainable Solutions. Paris, France: UNEP, Division of Technology Industry and Economics, Production and Consumption Branch
115. United Nations Environmental Programme, UNEP (2006) Design for sustainability: a practical approach for developing countries. United Nations Environmental Program, Paris
116. Van Gorp T (2012) Design for emotion. Morgan Kaufmann, Waltham, MA
117. Van Hinte E (1997) Eternally yours: visions on product endurance. 010 Publishers, Rotterdam
118. Van Kleef E, Van Trijp HCM, Luning P (2005) Consumer research in the early stages of new product development: a critical review of methods and techniques. Food Qual Prefer 16 (3):181–201
119. Von Hippel E (1988) The sources of innovation. Oxford University Press, New York and Oxford
120. Vezzoli C, Manzini E (2008b) 'Review: design for sustainable consumption and production systems', In Tukker A, Charter M, Stø E, Andersen MM Vezzoli C (eds) System innovation for sustainability 1. perspectives on radical changes to sustainable consumption and production. Greenleaf Publishing, Sheffield, UK
121. Vezzoli C (2003a) Systemic design for sustainability. In: Proceedings, Cumulus working paper, UIAH, Helsinki
122. Vezzoli C (2018) Design for Environmental Sustainability (London: Springer)
123. Wang B (ed) (1997) Integrated product, process and enterprise design. Chapman and Hall, London
124. Weidema B.P (2005) The integration of economic and social aspects in Life Cycle Impact Assessment, paper 2.-0 LCA consultants, Copenhagen, Denmark
125. Zaring O, Bartolomeo M, Eder P, Hopkinson P, Groenewegen P, James P, de Jong P, Nijhuis L, Scholl G, Slob A, Örninge M (2001) Creating eco-efficient producer services. Gothenburg Research Institute, Gothenburg
126. Zhang T, Dong H (2008) Human-centred design: an emergent conceptual model. Include 2009, Royal College of Art, London. April 8–10, 2009. Retrieved from http://bura.brunel.ac.uk/handle/2438/3472
127. Zoltowski CB, Oakes WC, Cardella ME (2012) Students' ways of experiencing human-centred design. J Eng Educ 101(1):28–59

Chapter 6
System Design For Sustainable Energy For All: A New Role For Designers

6.1 System Design for Sustainable Energy for All

We understood in the previous chapters that Sustainable Product-Service Systems (S.PSS) applied to Distributed Renewable Energy (DRE) represents a win-win opportunity to extend the access to sustainable energy to All. Indeed, this opens a new challenging role for designers, which claims for new knowledge-base and know-how, shortly defined as System Design for Sustainable Energy for All (SD4SEA). This role can be defined as follows:

> the design of a **Distributed Renewable Energy Sustainable Product-Service System**, able to fulfil the demand of sustainable energy of low- and middle-income people (All) - possibly including the supply of the Energy Using Products/Equipment - based on the design of innovative interactions of the stakeholders, in which economic and competitive interest of the providers, continuously seek after both socioethically and environmentally beneficial new solutions.

This SD4SEA role could be described by highlighting the main approaches and the related skills:

A. **'satisfaction-system' approach**: *design the energy access—possibly including the satisfaction of a particular demand (satisfaction unit)—and all its related products and services.*

B. **'stakeholder configuration' approach**: *design the interactions of the stakeholder of the energy access system—possibly including those related to a particular 'satisfaction unit'.*

C. **'system sustainability & energy 4all' approach**: *design such a stakeholder interaction (offer model) for economic and competitive reasons which continuously seek after both socioethical and environmentally beneficial new solutions, while/by providing Sustainable Energy for All.*

C. Vezzoli et al., *Designing Sustainable Energy for All*, Green Energy and Technology, https://doi.org/10.1007/978-3-319-70223-0_6

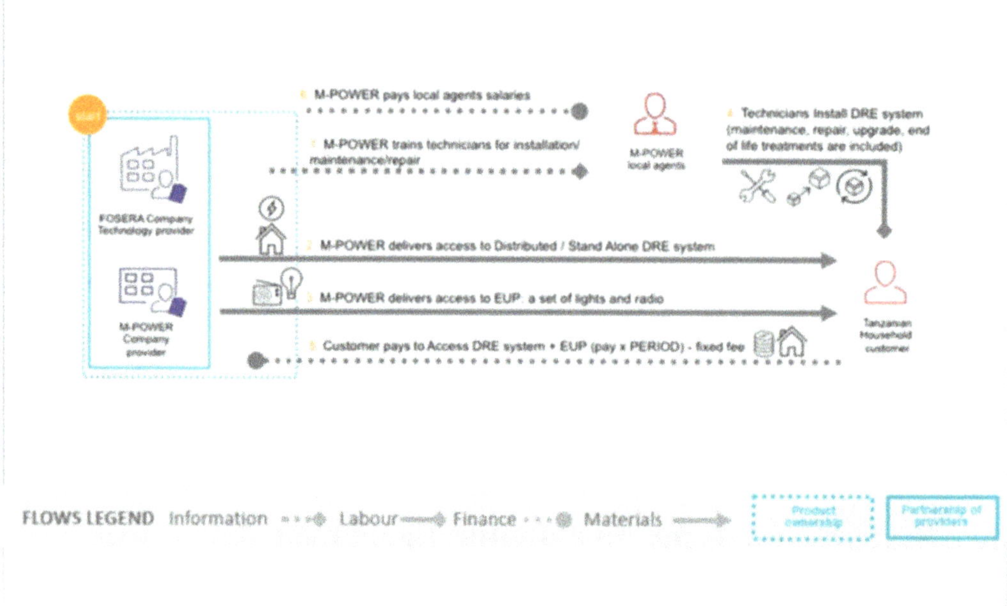

Fig. 6.1 Visualization of the System Map for an 'Appropriate Stakeholder Configuration Design'. *Source* designed by the Authors

It is important to highlight that SD4SEA is a new role for designers that derives and is declined to design DRE with its peculiar characteristics. This vision supposes the couple of 'Appropriate Technologies Design' with 'Appropriate Stakeholder Configuration Design', addressed to S.PSS&DRE (Fig. 6.1).

To clarify this concept, let us take a look at the System map (Fig. 6.2) a visualisation of the results of a stakeholder configuration design process and design tool. It is built up by a set of stakeholders and by a set of interactions in between them, namely material, financial and information flows.

However, as we have mentioned before, not all S.PPSs are sustainable. Even though good ideas and solutions may seem sustainable at the beginning, looking into the whole system may not. Because of this reason, criteria and guidelines are needed (as well as coherent support *methods* and *tools*) to orientate design towards eco-efficient and socioethical stakeholder interactions. Within the LeNSes project, a set of criteria and related guidelines have been developed and are presented together with some examples in the following chapter.

6.2 SE4A Design Criteria, Guidelines and Examples

The following set of six design criteria and related guidelines could be used by the designer to develop Distributed Renewable Energy as Sustainable Product-Service System.

Fig. 6.2 Indigo, Sub-Saharan Africa. *Source* www.azuri-technologies.com

First, the list of criteria is provided. Consequently, each criterion and related guideline is exemplified through case studies.

Criteria to develop Distributed Renewable Energy as Sustainable Product-Service Systems.

1. Complement the DRE offer with life cycle services (turnkey based);
2. Offer ownerless DRE systems as enabling platform;
3. Offer ownerless DRE systems with full services;
4. Add to DRE offer, the supply of ownerless Energy-Using Products and/or Energy-Using Equipment;
5. Delinked payment from pure watt consumption (affordable costs);
6. Optimise DRE systems configuration.

1. **Complement the DRE offer with life cycle services (turnkey based)**

This means to think about providing a business solution, which offers/sells to customers DRE systems (e.g. the energy generator, the storage or battery, the inverter and the wiring) complemented by different support services such as financial, design, installation, maintenance, repairing, upgrading and end-of-life treatment. The guidelines invite to design life cycle services that could be valuable in relation to the defined customer/s and unit of satisfaction.

1a. Complement the DRE offer, with financial services to support initial investment and installation costs, e.g. microcredit, crowdfunding and donation.
1b. Complement the DRE offer, with support services for the design and installation of its components, e.g. the generator, the storage, the inverter and the wiring.
1c. Complement the DRE offer, with support services during use, i.e. maintenance, repairing and upgrading of its components.
1d. Complement the DRE offer, with support services for the end-of-life treatment of its components.

1a. Example for complement DRE offer with financial services

Grameen Shakti/since 1996
Category: Solar Energy
Provider/s: Grameen Shakti
Customer: Households
Location: Bangladesh

As introduced (paragraph 4.4.2), Grameen Shakti offers Solar Home Systems (SHS) with a service package which includes end-user credit, installation, maintenance and repair, and take-back services. End users can benefit from a *financial service*, which allows them to purchase the SHS with microcredit services and repay the loan in 3–4 years. This means no initial investment cost for customer who becomes owner of the SHS with effective after-sale services included.

1b. Example for complement DRE offer with support services for the design and installation

Indigo/since 2012
Category: Solar Energy
Provider/s: Azuri Technologies
Customer: Households
Location: 11 countries around Sub-Saharan Africa

Indigo allows customers to purchase a Solar Home System (SHS) existing of a 2–5 W solar panel, battery, the charge controller, two LED lamps and a phone charge unit with cables, for only 10€. After the first payment, the SHS is installed by local dealer at the customer place, to use the SHS pays on a pay-as-you-go system: buying 1€ scratch card to access electricity for a week (eight hours of light each day and mobile phone charging) by inserting the code in the SHS charge controller. After 18 months, the purchase of scratch cards allows the system to be paid off and the customer can choose to either unlock his/her SHS or to upgrade to a larger model.

Indigo designs and produces the charge controller of the SHS, as key products to calculate energy expense and availability. The other components of the SHS come from other produces and are designed to meet the Lighting Global Quality Standards.

1c. Example for complement DRE offer with support services during use

Bboxx solar energy company/since 2010
Category: Solar Energy
Provider/s: Bboxx
Customer: Households
Location: Africa, Asia

Aside its offer introduced (paragraph 4.4.2), Bboxx has built up 45 shops across six countries in Africa and Asia, where it sells its own Solar Home Systems (SHS) and related appliances. The units are SMART and GSM enabled, and are remotely connected to a central database. Bboxx uses its platform, called 'SMART Solar', to monitor energy consumption and the performance of the systems. Customers pay a monthly fee (from 10 to 20 USD) depending on the size of the system and their chosen accessories. Installation and maintenance are included in a service fee and are done by Bboxx's local technicians. After complete repayment, the customer can go for a maintenance contract, which means he/she continues to get support and replacements for the unit, battery and panel. After around 3 years of payments, the customer owns the appliances. Analysis of data is used to optimise products and extend the life of the batteries, as such diminishing the frequency of replacement (Fig. 6.3).

2. **Offer ownerless DRE systems as enabling platform**

This means to think about providing a business solution, which offers to customers DRE systems, owned by the provider, as platforms that enable customers to operate on them to access to energy.

The guidelines invite to design enabling services that could be valuable in relation to the defined customer/s and unit of satisfaction.

2a. The energy supplier (existing or newly established) complements an ownerless offer of the DRE system—micro-generator eventually with some of accessories (storage, inverter, wiring, etc.) and/or the mini-grid—with training/information services to enable the customer to either design, instal, maintain, repair and/or upgrade one or more DRE components.

Fig. 6.3 Bboxx, Africa and Asia. *Source* www.bboxx.co.uk

2b. The micro-generator producer complements an ownerless offer of the micro-generator—eventually with its accessories and/or the mini-grid—with training/information services to enable the customer to either design, instal, maintain, repair and/or upgrade the micro-generator.

2c. The storage and/or the inverter, etc. producers complement an ownerless offer of their products with training/information services to enable the customer to either instal, maintain, repair and/or upgrade their products.

2d. A partnership composed by two or more stakeholders among the energy supplier, the micro-generators producer, the storages producer, the inverters producer, etc., complements an ownerless full package offer of their products with training/information services to enable the customer to either instal, maintain, repair and/or upgrade them.

2a. Example for complement an ownerless offer with training/ information services

Sunlabob solar energy/since 2000
Category: Solar Energy
Provider/s: Sunlabob, local committee
Customers: Inhabitants
Location: Laos

As introduced (see paragraph 4.4.2), Sunlabob leases a charging station with Energy-Using Products (EUP—e.g. solar lanterns) to an established village committee who rents the products to the individual households. Sunlabob supports the setting and training of a local committee which is responsible for setting prices, collecting rents and perform basic maintenance. To use the charging station, the committee pays around 1.70 € per month, without having the ownership of it. People can participate to the income-generating activities being part of the village committee, this increasing local competence and income (Fig. 6.4).

3. **Offer ownerless DRE systems with full services**

This means to think about providing a business solution, which offers the final satisfaction, i.e. the access to energy, and the customers neither own nor operate the DRE system. The guidelines invite to design full packages of services that could be valuable in relation to the defined customer/s and unit of satisfaction.

3a. The energy supplier (existing or newly established) complements the ownerless offer of the DRE system—micro-generator eventually with some of its accessories (storage, inverter, wiring, etc.) and/or the mini-grid—with the offer of one or more life cycle support services, i.e. installation, maintenance, repairing, upgrading and end-of-life treatment.

Fig. 6.4 Sunlabob, Laos. *Source* www.sunlabob.com

3b. The micro-generator producer complements the ownerless offer of the DRE system, with the offer of one or more life cycle support services, i.e. installation, maintenance, repairing, upgrading and/or end-of-life treatment.

3c. A partnership composed by two or more among the energy supplier, the micro-generator producer, the storages producer, the inverters producer, etc., complements the ownerless full package offer of their products, with one or more life cycle support services.

3a. Example for energy supplier complements an ownerless offer of the DRE system with the offer of one or more life cycle support services

OMC Power/since 2011
Category: Hydro/Solar/Wind/Hybrid Energy
Provider/s: OMC Power
Customer: Telecommunication companies and communities
Location: India

As introduced (paragraph 4.4.2), OMC Power offers energy solutions to telecommunication companies, through stand-alone power plants running on solar, wind and biogas. Telecommunication companies get the power plant installed on site and pay according to the energy they use (kWh). OMC Power retains the ownership of the energy system and provides operation and maintenance during the whole life cycle of the plant. The opportunity of having access to renewable and stable electricity increases reliability and continuity of companies in their work.

4. **Add to DRE offer, the supply of ownerless Energy-Using Products and/or Energy-Using Equipment**

This means to think of providing a business solution, which offers to customers, in addition to DRE systems offered (in one of previous three modalities), the supply of ownerless products that run on energy such as Energy-Using Products, e.g. light bulbs and radio, and/or Energy-Using Equipment, e.g. sewing machine and washing machine. The guidelines invite to offer Energy-Using Products/Equipment through S.PSS logic that could be valuable in relation to the defined unit of satisfaction.

4a. The energy supplier complements the offer of ownerless DRE system and its life cycle services, with the offer of Energy-Using Products and/or Energy-Using Equipment (ownerless and/or complemented with life cycle services).

4b. The micro-generator producer complements the offer of ownerless micro-generator and its life cycle services, with the offer of Energy-Using Products and/or Energy-Using Equipment.
4c. An Energy-Using Products or Energy-Using Equipment producer complements the offer of ownerless products and their life cycle services, with DRE system offer.
4d. A partnership composed by two or more stakeholders among the energy supplier, the micro-generators producer, the storages producer, the inverters producer, etc. and the Energy-Using Product producer, offer a full package of ownerless DRE system and Energy-Using Products or Energy-Using Equipment with their life cycle services.

4a. Example for energy supplier complement the offer of ownerless DRE system and life cycle services

Husk Power Systems (HPS)/since 2007
Category: Biomass Energy
Provider/s: Husk Power
Customer: Households and companies
Location: India

As introduced (paragraph 4.4.2), Husk Power System (HPS) provides energy solutions by installing biomass power plants and wiring villages to deliver electricity. The company retains ownership of the DRE plant and employs local agents for operation, maintenance and fee collection. In some villages with grid power, households and businesses choose to connect to the HPS supply because of its reliability and lower cost. HPS provides full medical benefits and retirement contributions for its full-time employees. Furthermore, farmers can earn an income from the sale of rice husks, and some residents have been trained to do maintenance and operation of the plant creating new income-generating activities (Fig. 6.5).

5. **Delink payment from pure watt consumption (affordable costs)**

This means to think of providing a business solution at affordable costs offering a type of payment dissociated from the energy consumption, e.g. customers pay either per demand, time or use/satisfaction, and the availability of energy depends on the maximum capacity of the DRE system installed. The guidelines invite to choose a payment modality that could be valuable in relation to the defined customer/s and unit of satisfaction.

Fig. 6.5 Husk Power Systems, India. *Source* www.huskpowersystems.com

5a. Offer pay *x* period, i.e. the cost is daily, weekly, monthly or yearly fixed.
5b. Offer pay *x* time of access to energy, i.e. the cost is fixed per minutes/seconds of access to energy.
5c. Offer pay *x* use/satisfaction unit of (energy-using) product, i.e. the cost is fixed per product performance (e.g. km for a vehicle, washing cycles for washing machine).
5d. Offer payment based on hybrid pay *x* period, pay *x* time and pay *x* use modalities.
5e. Offer payment with the support of additional financial support from public administrations/entities.

5a. Example for DRE offer as pay per period

OFF-GRID Electric/since 2012
Category: Solar Energy
Provider/s: M-POWER
Customer: Households
Location: Tanzania

As introduced (see paragraph 4.4.2), M-POWER offers to Tanzania rural people Solar Home Systems (SHS) (Solar panel + Storage + Wires) and the related Energy-Using Products (EUP) (two lights + phone charger) as a pay per period with a daily/weekly/monthly fee. M-POWER retains the ownership of SHS and EUPs including their maintenance and repair.

5c. Example of DRE offer as pay per use/satisfaction unit

Solar-Powered Café/since 2001
Category: Solar Energy
Provider/s: Solar Charge
Customer: Inhabitants
Location: South Africa

The Solar-Powered Café pilot project offers a solar-powered connection centre and charging point, bringing low-cost access to IT services. Ownership of the connection centre and charging point (and of all the included Energy-Using Products) is retained by Solar Charge. The customer pays per use with three different offers at same price: one internet access, one IT service and one phone charging. The connection centre has a highly trained administrator to manage any problems that may arise (Fig. 6.6).

6. **Optimise DRE systems configuration**

This means to think of providing a business solution with the best-optimised configuration for the DRE system according to the context conditions. In other words, understand whether providing distributed or decentralised stand-alone systems for off-grid contexts or creating a distributed or decentralised mini-grid to share the energy surplus. The guidelines invite to optimise the DRE systems configuration in relation to the defined customer/s, unit of satisfaction and context of use.

6a. Offer stand-alone Distributed Renewable Energy (DRE) systems for homes or business sites (especially to off-grid and isolated sites).
6b. Offer local mini-grids connecting DRE systems, to enable local energy surpluses sharing (especially for context with nearby energy-consuming units).
6c. Offer Decentralised Renewable Energy stations as charging spot or energy use services spot with EUP/EUE for local communities.
6d. Offer Decentralised Renewable Energy systems to enable local supply of energy throughout a mini-grid for homes and/or business sites.
6e. Offer DRE system with connection to main-grid, enabling homes, small business and local mini-grids the selling/purchasing of energy.

Fig. 6.6 Solar-Powered Café, South Africa. *Source* www.kutengatechnology.com

6a. Example for DRE offer as stand-alone DRE system

Domestic Biogas/since 2007
Category: Biogas Energy
Provider/s: Biogas Sector Partnership, customer, partnerships with private companies
Customer: Households
Location: Nepal

Biogas Sector Partnership instals biogas plants as distributed stand-alone DRE systems in households, providing biogas for cooking and lighting. A plant costs between 350 and 450€; about one-third of this is paid in-kind, through the family providing labour and materials. The remaining is paid usually in 18 months, with opportunity of micro-financing plans. Customers are trained for minor repairs and operations on plants; a 3-year guarantee period is included (Fig. 6.7).

Fig. 6.7 Domestic Biogas, Nepal. *Source* www.ashden.org

6c. Example for DRE offer as decentralised DRE station

Solar Transition/Since 2011
Category: Solar Energy
Provider/s: Ikisaya Energy Group (Cooperative-Based Organisation)
Customer: Community
Location: Kenya

Solar transition, settled in Ikysawa village in Kenya, is a village decentralised DRE station that provides renewable energy for a range of daily services: lantern and battery charging and renting, charging of mobile phones, IT services (typing, printing and photocopying), television and video shows. The decentralised DRE station is provided with the hardware to generate solar energy (Solar panel + Storage + Wires) and a configurable series of Energy-Using Equipment (EUE). Solar transition recharging station is owned and managed by the community itself who becomes local entrepreneur with competences on maintenance and repair. Customers have first to pay an initial membership fee; so forth they pay only for each service they use, as a pay per use payment. The opportunity to access several services related to energy (e.g. print, computer use) facilitates local communication with activities outside from the villages, and families can socialise in the common space.

6d. Example for DRE offer as decentralised renewable energy system throughout MINIGRID

Micro-hydro grid/since 1996
Category: Hydropower Energy
Provider/s: CRELUZ (Cooperativa de Energia e Desenvolvimento Rural do Médio Uruguai Ltda)
Customer: Community
Location: Brazil

The project provides decentralised renewable energy plants, in the configuration of mini-hydropower plants, connected through local mini-grid (already existing), generating part of the community electricity needs. Customers pay the electricity used in their home connected to the mini-grid in the various payment points available. Local operators have been trained on the technical aspects of the hydro plant, as part of the educational project to make people aware of power generation. Maintenance and repair is done by CRELUZ, and emergency phone service is guaranteed 24 h.

Part III
Method and Tools for SD4SEA

Chapter 7
Method and Tools for System Design for Sustainable Energy for All

7.1 Method for System Design for Sustainable Energy for All

The method developed within the LeNSes project is called Method for System Design for Sustainable Energy for All (MSD4SEA). It came out as one of the results of the project, but it is based on other methods and tools developed formerly under other EU funded researches.

The method aims to support and orient the entire process of system innovation development towards Sustainable Energy for All. It is conceived for designers and companies but is also appropriate for public institutions and NGOs. It can be used by an individual designer or by a wider design team. In all cases special attention is given to facilitating both within the organisation itself (between people from different disciplinary backgrounds) and outside, bringing different socio-economic actors and end-users into play co-designing processes.

The method is organised in stages, processes and sub-processes. It is characterised by a flexible modular structure so that it can easily be adapted to the specific needs of designers/companies and to diverse design contexts and conditions. Its modular structure is of interest in the following:

- Procedural stages: all the stages can be used or certain stages can be selected according to the requirements of the project;
- Tools to use: the method is accompanied by a series of tools (many of them elaborated within the same LeNSes project). It is possible to select which of these to use during the design process;
- Integration of other tools and activities: the method is structured in such a way as to allow the integration of design tools that have not been specifically developed for it. It is also possible to modify existing activities or add new ones according to the requirements of the design project.

C. Vezzoli et al., *Designing Sustainable Energy for All*,
Green Energy and Technology, https://doi.org/10.1007/978-3-319-70223-0_7

The basic structure of method consists of four main stages.

- Strategic analysis;
- Exploring opportunities;
- Designing system concepts;
- Designing (and engineering) a system.

A further stage is added, across the others, to draw up documents to report on the sustainability characteristics of the solution designed, namely:

- Communication.

The following Table 7.1 shows the aims, the processes and the tools for each stage of the method.

Table 7.1 Stages, aims, processes and tools for each stage of the method for SD4SEA

Aims	Processes	Tools
Strategic analysis (SA)		
To obtain information to facilitate the generation of S.PSS applied to DRE systems	Analyse project proposers and reference context and general macro-trends	– Innovation diagram for S.PSS and DRE – Energy System map – S.PSS + DRE innovation map – Strategic analysis (SA) template – MiniDOC – SWOT matrix
	Analyse sustainability of existing system and set priorities for the design intervention	– Strategic analysis (SA) template – S.PSS + DRE innovation map – Sustainability design orienting (SDO) toolkit
	Analyse access to energy in the context of reference	– Strategic analysis (SA) template – Resources assessment software – Sustainability design orienting (SDO) toolkit
	Analyse sustainable best practices	– Energy system map – S.PSS&DRE case study format – Sustainability design orienting (SDO) toolkit – MiniDOC

(continued)

Table 7.1 (continued)

Aims	Processes	Tools
Exploring opportunities		
To make a 'catalogue' of promising opportunities towards S.PSS applied to DRE	Generate sustainability-oriented ideas at system/stakeholder level	– Offering diagram – Sustainability design orienting (SDO) toolkit – Satisfaction system map
	Generate DRE oriented ideas at system/stakeholder level	– Sustainability design orienting scenario for S.PSS&DRE – Sustainable energy for all idea tables (and cards) – S.PSS + DRE design framework and cards
	Outline a sustainable design orienting scenario	– Sustainability design orienting scenario – S.PSS + DRE innovation map
Design system concepts		
To determine one or more system concepts oriented towards S.PSS applied to DRE	Select clusters and single ideas (environmental, socioethical, DRE-oriented)	– Innovation diagram for S.PSS and DRE
	Develop system concept/s	– Energy system map – PSS + DRE Design framework and cards – Estimator of DRE (E.DRE) – Concept description form for S.PSS and DRE – Stakeholder's motivation and sustainability table – Offering diagram – Interaction table – Interaction storyboard – System concept audiovisual
	Environmental, socioethical, and economic assessment of system concept/s	– Sustainability design orienting (SDO) toolkit – Sustainability interaction story-spot
	Evaluate the system concept/s	– Stakeholder's motivation and sustainability table

(continued)

Table 7.1 (continued)

Aims	Processes	Tools
Design system details		
To develop the most promising system concept into the detailed version ready for implementation	Detail the system	– Energy system map – Offering diagram – Interaction table – Interaction storyboard – Stakeholder's motivation and sustainability table – Solution element brief – Business plan
	Environmental, socioethical, and economic assessment of DRE system	– Sustainability design orienting (SDO) toolkit
	Present/discuss the system developed, e.g. outline main activities characteristics, actors	– Sustainability interaction story-spot – Animatic
Communication		
To communicate (internally/externally) the general and (above all) sustainable characteristics of the system designed	Draw up the documentation for (internal) communication	– Sustainability design orienting (SDO) toolkit – MiniDOC
	Draw up the documentation for (external) communication	– Animatic – Energy system map – Offering diagram – Interaction story-spot – Sustainability design orienting (SDO) toolkit

Source designed by the Authors

The following sections present each stage describing its component processes. Attention is paid to sustainability-orienting processes.

Strategic Analysis

The aim of the first part of the method is to collect and process all the background information necessary to the generation of a set of potentially sustainable ideas. The objective is twofold: on the one hand, to understand the existing situation and find out more about the project proposers, the socio-economic context in which they operate and the dynamics (socio-economic, technological and cultural macro-trends) that influence that context; on the other hand, to process information by which to steer the designing process towards the generation of promising solutions, favouring sustainable energy access to All. The processes are outlined below.

Analyse project promoters and outline the intervention context

Given that the project proposers may be companies, public institutions, NGOs, research centres, or a mix of these, the aim of this activity is first and foremost to define the scope of the design intervention, or rather the *satisfaction unit* to be met (e.g. move around the city for working purposes or have clean clothes). At this point, the characteristics of the project proposers are examined carefully: their 'mission', their main areas of expertise, their strength and weaknesses, opportunities and threats, in relation to the area of intervention. In addition, particularly, if the proposer is a company, the value chain will be analysed to understand how this is structured, what actors come into play, what problems (environmental, socioethical and economic) may be met.

Key questions:

- What is the demand/satisfaction unit to be met?
- What are the key areas of expertise of the project promoters?
- What are their main strengths and weaknesses?
- Who are the main actors? What is the relationship between them?
- What are the main environmental, socioethical and economic problems associated with the value chain?
- What is the value for the customer?

Analysing the context of reference

The aim of this activity is to analyse the context, or rather the sociotechnical regime, of which the innovation will become a part. First, the structure of the production and consumption system (the scope of intervention) is analysed: what actors come into play (companies, institutions, NGOs, consumers, etc.) and what the relationships are between them, as well as what specific dynamics (technological, cultural, economic and regulatory) characterise the system itself. Special attention is also paid to current and potential competitors (analysing their characteristics and offers) and to customers (analysing their needs).

Key questions:

- How is the entire production and consumption chain structured in relation to the scope of intervention (satisfaction unit)? Who are the main actors (public and private) and their respective interests?
- What are the technological, cultural and regulatory dynamics influencing, or of potential influence, the characteristics of the production and consumption chain?
- Who are the main competitors? What are their offers and how do these differ from those of the project proposers?
- Who are the potential customers? What are their needs? Are their needs satisfied?

Analysing the carrying structure of the system
The aim of this activity is to identify and analyse the general macro-trends (social, economic and technological) that lie behind the reference context. It is important to understand these in order to understand what potentially influences the context (or sociotechnical regime) that will be the object of the intervention.

Key question:

- What are the main social, economic and technological macro-trends?
- How may these influence the reference context and consequently the design options?

Analysing cases of sustainable energy access
The aim of this activity is to analyse in detail cases of excellence that could act as a stimulus during the generation of ideas. The result will be a document summarising the offer for each case of excellence, the interactions with the user, the offer producers and providers, and its sustainability characteristics.

Key questions:

- What is the offer, in terms of products and services? How does the user interact with the offer?
- Who are the actors in the offer system? What are their intentions?
- What are the environmental, socioethical and economic benefits?

Analysing the context energy access
The aim of this activity is to analyse the access to (renewable and non-renewable) energy sources within the context where the existing offer is given.

Key questions:

- How is energy delivered (Country/region energy plan/local policies for energy access)?
  - Within the country/region?
  - Within the specific design context?

- Which are the main energy sources used? (Renewable/non-renewable)
  - Within the country/region?
  - Within the specific design context?
- Why some areas donot have energy? How they currently supply energy need?
- If energy provision is not guaranteed/legal, how the users currently supply energy need in the design context?
  - In the country/region?
  - In the specific design context?
- What is the average electricity consumption of households per capita?
  - In the country/region?
  - In the specific design context?
- Who, between men and women, has the control on the energy use in the specific design context?
- Which is the availability and capacity of renewable energy resources)?
  - In the country/region?
  - In the specific design context?

Analysing sustainability of existing system and determine priorities for the design intervention in view of sustainable energy solutions and sustainability more in general
The aim of this activity is to analyse the existing energy system in the design context from environmental, socioethical and economic point of view in order to identify the design priorities (in other words, to understand where it is most important to intervene in order to reduce the environmental, socioethical and economic impact of the existing energy system). This operation is fundamental to steering the design process towards the solutions that are the most able to foster Sustainable Energy for All. The result will be a document summarising the energy system analysis and its environmental, socioethical design priorities.

Key questions:

- What is the situation in the design context regarding the existing energy system and its environmental, socioethical and economic sustainability?
- What are the design priorities for each dimension of sustainability?

Exploring opportunities
The aim of the second stage is to identify possible orientations for the development of promising solutions. This takes place through a participatory process, whereby the various actors generate ideas.

It must be stressed that the aim of this stage is not to come up with incremental improvements at product level, but rather to come up with possible innovations at system level, characterised by radical improvements from an environmental, socioethical and economic point of view.

The specific aim is therefore to use all the information collected and processed during the previous stage to outline a 'catalogue' of promising strategic S.PSS applied to DRE opportunities.

Generate sustainability-oriented ideas
On the basis of the information previously acquired, a set of potentially sustainable ideas is generated through an idea-generating workshop. It must be made clear that the idea generation must be orientated towards satisfying a specific satisfaction unit. In this sense, particular attention is paid to coming up with system level ideas, i.e. ideas regarding the configuration of actors able to produce/deliver that offer (satisfaction unit); and the products and services that constitute the offer. Special design guidelines have been drawn up to steer idea generation towards sustainable system solutions. It is also useful to have a collection of cases of excellence available as a further stimulus, and a map of the actors who may potentially become part of the *satisfaction system*. The result of this process will be a document listing the satisfaction unit and a set of system ideas with their environmental, socioethical and economic sustainability characteristics.

Key questions:

- What is the satisfaction unit to be met by design?
- Who are the actors who may potentially be involved in the *satisfaction system*?
- What potential product and service systems are capable of bringing radical improvements (from an environmental, socioethical and economic point of view)? What actor system will be able to produce and deliver such an offer?

Generate Energy for All-oriented ideas
The aim of this process is to orientate system idea generation design process towards promising Sustainable Energy for All solutions. Generally, ideas are generated through workshops, starting with the definition of the energy satisfaction unit to be met by design.

Specify the Sustainable Energy for All design-oriented scenario
The aim of this stage is to specify in relation to the context, the providers and the satisfaction unit, the Sustainable Energy for All design orienting scenario, the scenario is composed of a set of visions, or better, possible promising Sustainable Energy for All design orientations.

The aim of this process is to select, map and cluster most promising ideas previously generated and place them in the Innovation Diagram for S.PSS & DRE tool, then generate new ideas to move from one polarity to another one generating further promising ideas.

System Concept Design
The aim of this stage is to select the most promising clusters and single ideas and design one or more system concepts oriented towards S.PSS applied to DRE solutions.

Selecting clusters of ideas and/or single ideas
The most promising ideas (environmental, socioethical and DRE oriented) are selected and combined through a participatory process, possibly supported by purposefully designed tools. Each of these combinations will then be developed into a system concept.

Key questions:

- Which ideas are the most promising from an economic point of view and in terms of provider's competences and customer acceptability?
- Which ideas are most promising from an environmental and socioethical point of view?

System Concepts development
One or more system concepts will emerge from the combinations of ideas previously singled out. The following elements are then defined for each of these system concepts: the set of products and services that make up the offer and the functions it fulfills; the actor system (primary and secondary) that produces and delivers the offer; and the interaction between various stakeholders of the satisfaction system.

Key questions:

- What products and services make up the offer? What functions does it fulfill? What is the value perceived by the user? How does the customer interact with the offer system?
- Who are the socio-economic actors of the system and what are their interactions? Which are the principal and which the secondary actors?

DRE System Concept Design
The aim of this process is to select the most appropriate renewable energy resource available in the context in which will be implemented the design solution and to estimate, according to the user-energy need, the size of the DRE.

Environmental, socioethical and economic assessment
The aim of this process is to assess the potential improvements that the system concepts could generate from an environmental, socioethical and economic point of view. This process is fundamental in order to understand whether there are still any unresolved critical points and also, if more than one concept has been developed, to decide which one is more promising. The result will be a description, for each concept, of the potential improvements offered (for every criterion of each sustainability dimension); a visualisation of these improvements by means of a radar diagram; and a visualisation of the interactions that illustrate improvements.

Key questions:

- What are the potential environmental, socioethical and economic improvements that the system concept can generate?

- Does the system concept have any critical points from an environmental, socioethical and/or economic point of view? Do any of its elements need redesigning?

System Design and Engineering

The aim of this stage is to itemise the specific requirements of the system concept to enable its implementation.

The processes connected to this stage are described below.

Detailed system design (executive level)

The aim of this activity is to develop the system concept in detail, defining: the set of products and services that make up the offer; all the actors (both primary and secondary) involved in the system together with their roles; all the interactions between actors including the customer that occur during delivery of the offer; all the elements (both material and non-material) required for delivery of the offer and who will design/produce/deliver them.

Key questions:

- What products and services make up the system? What are the primary and secondary functions delivered? What value is perceived by the customer? How does the customer interact with the offer system?
- Who are the actors (both primary and secondary) that take part in the system? What kind of interactions (partnerships, agreements) do they have? What are their respective roles and interactions in delivering the offer?
- What material and non-material elements are required to deliver the offer?

Environmental, socioethical and economic assessment

The aim of this activity is to assess more accurately the environmental, socioethical and economic benefits that the system innovations will produce once implemented. The result will be a more detailed description of the potential improvements for each project (for every criterion of each sustainability dimension), a visualisation of these improvements by means of a radar diagram, and a visualisation of interactions that illustrate the improvements.

Key questions:

- What environmental, socioethical and economic improvements can be expected from the implementation of the system innovations designed?

Communication

The communication stage aims to communicate the general characteristics of the solution designed, and above all those regarding sustainability, to the outside world.

The basic aim is to provide a document indicating:

- The general characteristics of the product-service system. The elements that make up the system innovation are described: the set of products and services that the offer consists of; the primary and secondary actors involved in the system and their respective roles and interactions; and the interactions between the actors and customer

- The sustainability characteristics of the product-service system. The potential improvements (from an environmental, socioethical and economic point of view) to be gained from the implementation of the solution are shown, with an indication of the elements of the system that will deliver these improvements.

7.2 SD4SEA Tools

The method includes not only a series of existing or adapted tools but also new tools designed, implemented and tested specifically to design S.PSS applied to DRE. These tools are listed below and will be described in this chapter:

- Sustainability Design Orienting Scenario on S.PSS&DRE;
- Strategic Analysis SA template;
- Sustainable Energy for All Idea Tables and Cards;
- E.DRE—estimator for distributed renewable energy;
- S.PSS + DRE Innovation Map;
- S.PSS + DRE Design Framework and Cards;
- Energy System Map;
- Innovation Diagram for S.PSS&DRE;
- Concept Description Form for S.PSS&DRE.

The description of the other following tools for S.PSS design could be found in the tool section of www.lenses.polimi.it:

- Satisfaction System Map;
- SDO toolkit;
- Interaction table (and storyboard);
- Offering Diagram.

The design tools will be described according to their aims, what they consist of, how to use them, integration in the design process, their results, their availability and required resources.

7.2.1 *Sustainability Design Orienting Scenario (SDOS) on S.PSS&DRE*

Aims

Design Orienting Scenario [11], a tool to inspire and inform designers towards possible futures on specific topics, has been adapted [1, 12] to Sustainable Product-Service System (S.PSS) applied to Distributed Renewable Energy (DRE). The tool, (from now on) Scenario presents four visions narrated as interactive videos accessible through a navigator file. The Scenario is a tool to inspire designers and stakeholders to design radically new social, economic and technical solutions and as co-design strategic conversations and facilitating creative processes among different actors (Fig. 7.1).

What it consists of

The tool allows to watch the videos to inspire towards Sustainable Product-Service System (S.PSS) applied to Distributed Renewable Energy (DRE) solutions. The tool presents four visions within a polarity diagram of two axes. The horizontal axis defines who is the customer of the narration final user (B2C), or as small entrepreneur/small business (B2B). The vertical one defines the offer: a Distributed Renewable Energy generator (e.g. solar panel system plus its appliances such as storage, inverter, wires, etc.), or the sum of both the Distributed Renewable Energy generator and the related Energy-Using Products or Energy-Using Equipment (e.g. phone and television are Energy-Using Products; woodworking machine, sewing machine are Energy-Using Equipment). Each vision is presented through one short video (around 90 s) that shows peculiar narration, highlighting the key points of the vision (e.g. stakeholder interactions, ownership. Three sub-videos (around 30 s each) help to achieve the understanding of a wider range of opportunities than presented in the video of the vision; these three sub-video show: all the possible offer and the related payment modality; (2) all the possible stakeholders that can be involved and their possible interactions; (3) all possible sustainability benefits (environmental, socioethical and economic).

How to use the tool

The Scenario requires the use of a slideshow software (e.g. Open Office PowerPoint). Each video and sub-video can be watched separately or a central button is available to run the whole videos as one. The suggestion is to watch a main video first and after the related sub-videos, then, the second main video and so on.

Integrating the tool into the design process

The Scenario can be used during the *Exploring Opportunities.*

Exploring opportunities

It can be used to inspire and inform designers and actors involved towards possible visions of Sustainable Product-Service System (S.PSS) applied to Distributed Renewable Energy (DRE), and to get new inspirations during the process.

Results

The result is a set of ideas favouring creative processes and co-design activities towards concepts of Sustainable Product-Service System (S.PSS) applied to Distributed Renewable Energy (DRE).

Tool availability and required resources

The tool is available for free download at www.lenses.polimi.it. The tool has been designed to be used in workshops and co-design sessions, therefore a projector is preferable. The time required to visualise all videos is approximately 15 min.

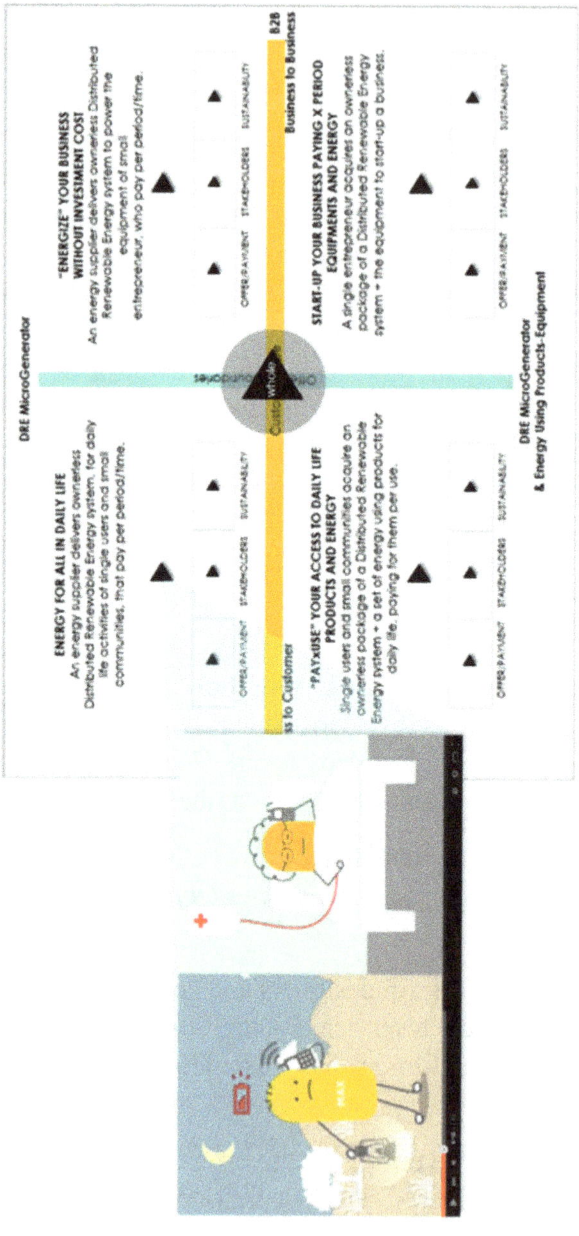

Fig. 7.1 Sustainability design orienting scenario on S.PSS and DRE. *Source* designed by the Authors

7.2.2 Strategic Analysis (SA) Template

Aims

Strategic Analysis (SA) template, a tool to collect and process the background information necessary to the generation of a set of potentially sustainable solutions. On the one hand, it aims to understand the existing situation and find out more about the existing proposers, the socio-economic context in which they operate and the dynamics (socio-economic, technological and cultural macro-trends) that influence that context; on the other hand, it aims to process information by which to steer the designing process towards the generation of promising sustainable solutions (Fig. 7.2).

What it consists of

The tool is an editable template based on five sections:

A. Design brief;
B. The context;
C. Existing system;
D. Qualitative sustainability evaluation of existing system;
E. Access to energy.

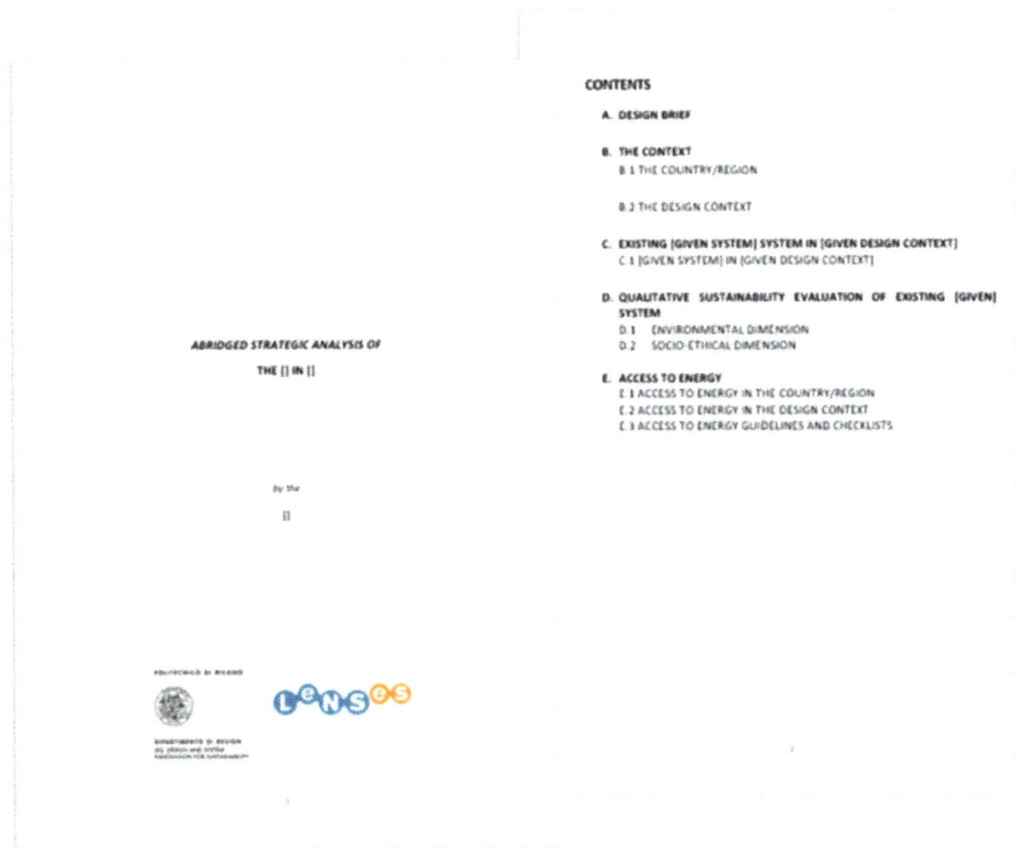
ABRIDGED STRATEGIC ANALYSIS OF

THE [] IN []

by the

[]

CONTENTS

A. DESIGN BRIEF

B. THE CONTEXT
B.1 THE COUNTRY/REGION
B.2 THE DESIGN CONTEXT

C. EXISTING [GIVEN SYSTEM] SYSTEM IN [GIVEN DESIGN CONTEXT]
C.1 [GIVEN SYSTEM] IN [GIVEN DESIGN CONTEXT]

D. QUALITATIVE SUSTAINABILITY EVALUATION OF EXISTING [GIVEN] SYSTEM
D.1 ENVIRONMENTAL DIMENSION
D.2 SOCIO-ETHICAL DIMENSION

E. ACCESS TO ENERGY
E.1 ACCESS TO ENERGY IN THE COUNTRY/REGION
E.2 ACCESS TO ENERGY IN THE DESIGN CONTEXT
E.3 ACCESS TO ENERGY GUIDELINES AND CHECKLISTS

Fig. 7.2 Strategic analysis (SA) template. *Source* designed by the Authors

For each section, a set of subsections with questions and/or guideline is available to support its completion.

How to use the tool
The Strategic Analysis (SA) template can be printed or edited in its digital version. Each section and sub-section can be filled separately and according to the aim of the activity.

Integrating the tool into the design process
The Strategic Analysis (SA) template can be used during the *Strategic Analysis*, aiming to collect preliminary information and (if needed) setting the bases for the design activity.

Results
The result is a collection of information about design brief, context and the related access to energy, existing system as well as its sustainability (environmental socioethical, economic).

Tool availability and required resources
The tool is available for a free download at www.lenses.polimi.it. To be used in its digital version, an editing software is needed (e.g. Open Office Word). The time required could last from hours to days, according to the detail of the information.

7.2.3 Sustainable Energy for All Idea Tables and Cards

Aims
Sustainable Energy for All Idea Tables [13], structured on the SDO toolkit,[1] it is a tool to generate ideas for S.PSS applied to DRE solutions, it is based on six idea tables with guidelines. To the guidelines of each table are connected 15 case studies to be used as examples. For each of the case study, a card has been developed. The Sustainable Energy for All Idea Tables is presented as a co-design tool to generate (sustainable) ideas facilitating the creation process (Figs. 7.3, 7.4 and 7.5).

What it consists of
The tool allows the generation of (sustainable) ideas for S.PSS applied to DRE solutions. Six idea tables with criteria and guidelines are available to orientate the design process. Fifteen case studies and cards can be used as supportive examples associated to the guidelines.

Each table refers to a *criterion* (and includes specific guidelines) to design (sustainable) ideas for an S.PSS applied to DRE concept. The criteria are the following and they are described with their guidelines in Chap. 5:

---

[1]The SDO Toolikt has been adapted to the new criteria and guidelines for sustainable energy for All. The SDO toolkit was developed by Carlo Vezzoli and Ursula Tischner within the MEPSS EU 5th FP, Growth project.

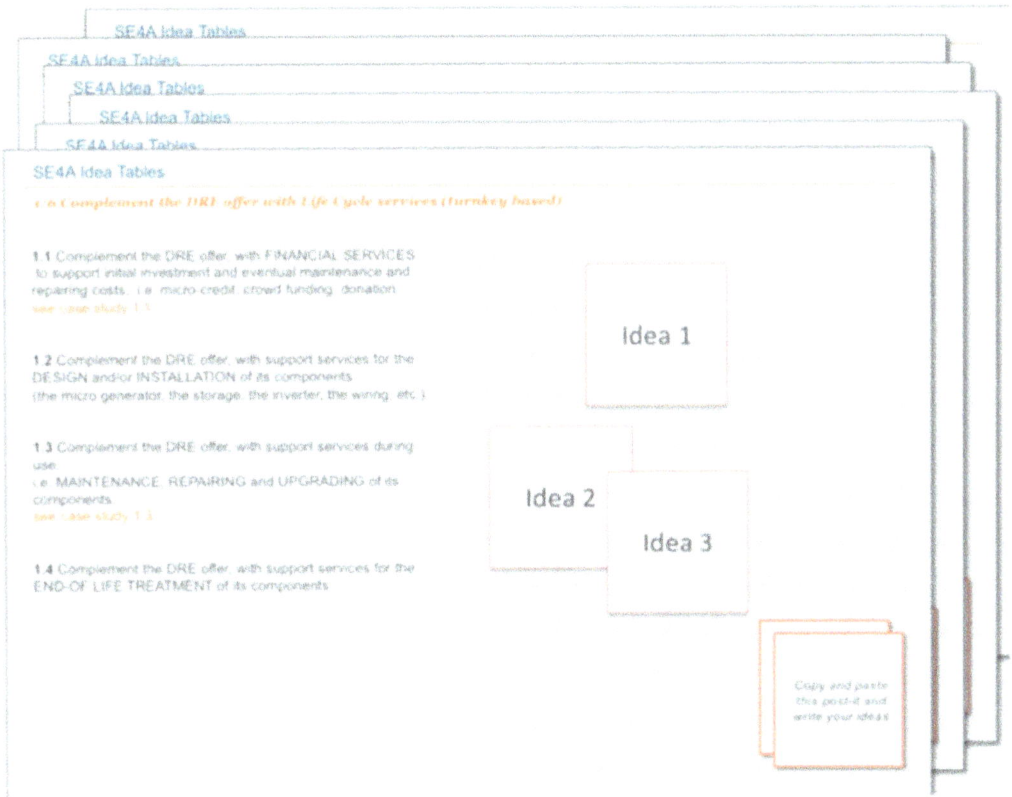

Fig. 7.3 Printable sustainable energy for all idea tables. *Source* designed by the Authors

1. Complement the DRE offer with Life Cycle services (turnkey based);
2. Offer ownerless DRE systems as enabling platform;
3. Offer ownerless DRE systems with full services;
4. Add to DRE offer, the supply of ownerless Energy Using;
5. Delinked payment from pure watt consumption (affordable costs);
6. Optimise DRE systems configuration.

How to use the tool

The Sustainable Energy for All Idea Tables tool could be used with two modalities.

- a slideshow software (e.g. Microsoft PowerPoint, or the equivalent in Open Office) or, in the case it is printed, it will require post-it and pens;
- the SDO toolkit software in the dimension of *Sustainable Energy for All.*

Use of the idea tables

Each table needs to be used singularly and presents a series of guidelines which are suggestions to orient the design of (sustainable) ideas in relation to a specific offer. Aside from the guidelines, an empty space is left to post ideas. After reading the guidelines, it is possible to use the post-it (digital or in paper) to write ideas. As general rules: no ideas are wrong; there is not a compulsory number of ideas to be written; the ideas need to be at system-stakeholders' interaction level and not at product level, e.g. offer the use of a bike (sharing) with a payment based on time of use to bring kids to school, but not a bike itself.

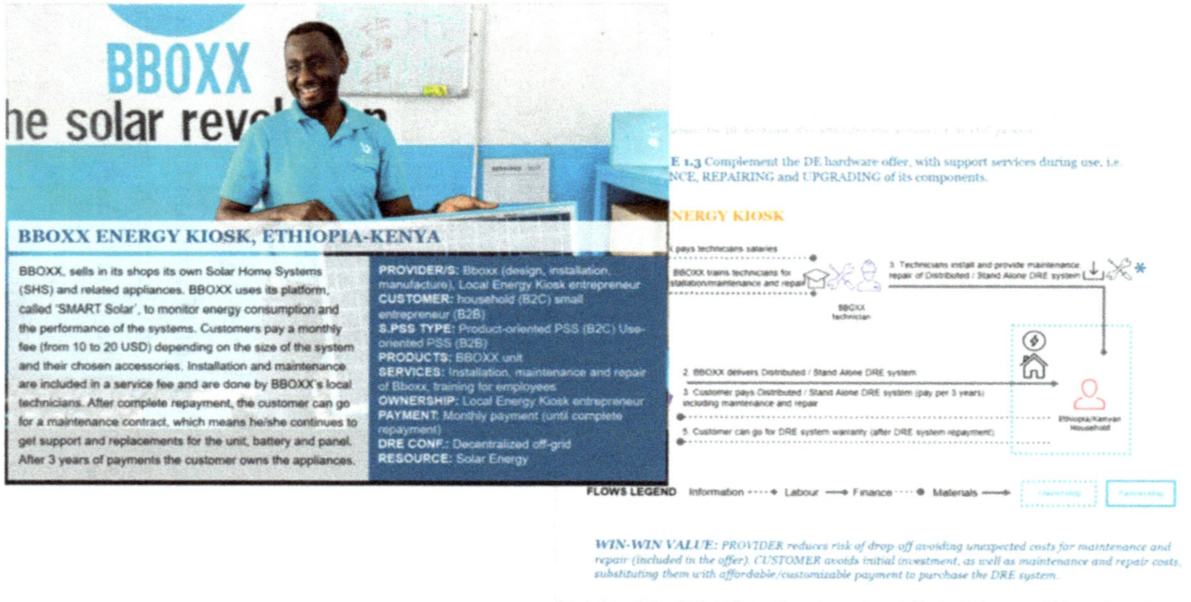

Fig. 7.4 Cards (with case studies) to complement the sustainable energy for all idea tables. *Source* designed by the Authors

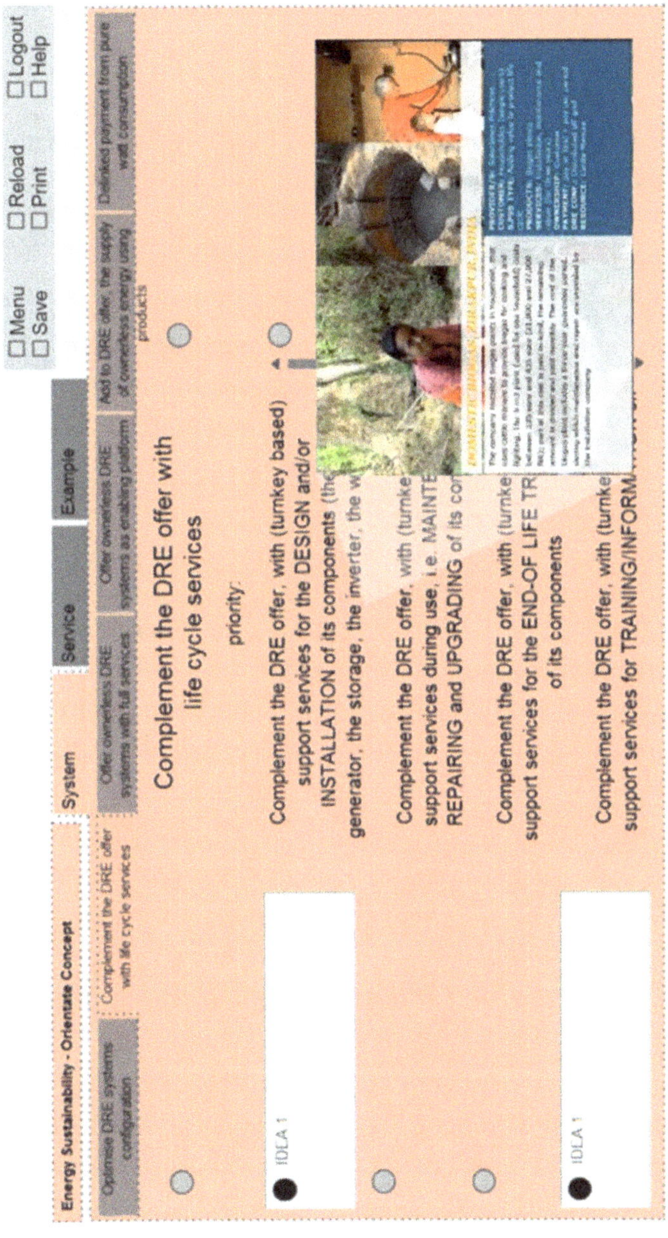

Fig. 7.5 Digital version of sustainable energy for all idea tables and cards (with case studies). *Source* designed by the Authors

Use of the case study either as online access or as cards
Each case study represents an existing case of S.PSS applied to DRE in relation to a specific guideline. Each card is made of a short description with the key information: customer, provider, type of S.PSS, offered products (and related ownership), offered services (and related provider), what is paid, DRE source, DRE system configuration (front of the card) and a visualisation of the stakeholder's interactions through an Energy System Map,[2] where the interaction representing the guideline is highlighted.

Integrating the tool into the design process
The Sustainable Energy for All Idea Tables and examples are used in the *Exploring opportunities* stage to support the generation of (sustainable) ideas towards S.PSS applied to DRE solutions.

Results
The results are new sustainable ideas (written in the post-it) of Sustainable Product-Service System (S.PSS) applied to Distributed Renewable Energy (DRE). The most promising ideas are transferred into the Innovation Diagram for S. PSS&DRE to generate the concept (more about in paragraph 2.6.7, where the tool is presented).

Tool availability and required resources
The tool is available for a free download and in copy-left at www.lenses.polimi.it. The tool has been designed to be used in workshops sessions, therefore is good to work on it collectively, though it could be used even by one person only. It is available both with digital version which could be used through a pc with or without a projector (suggested if the group is composed by more than 3–4 person) or as printable one (suggested to be printed as A3–A2). The case study cards are available in digital and printable version, the suggestion is to print them to facilitate the exchanges between the group. The time required is approximately 60 min (10 for each idea table).

7.2.4 E.DRE—Estimator for Distributed Renewable Energy

Aims
The tool [13] is developed to support the design of Distributed Renewable Energy (DRE) systems, as well as to guide the evaluation of the energy demand and need of the designed system concept, and to assess the best system configuration and estimate the energy production potential (Fig. 7.6).

[2] *Energy System Map tool has been developed in the LeNSes project (see paragraph 6.6).*

Fig. 7.6 S.PSS and DRE estimator of DRE load/need and production potential. *Source* designed by the Authors

What it consists of
The tool is composed of six main worksheets (in one excel file):

1. Worksheet for *energy load/need* and *energy production potential*, it summarises the energy load/need to satisfy the system appliances, and compares such data with the table that summarise the energy production potential of the DRE system (existing or designed);
2. Worksheet for *Energy-Using Product (EUP) consumption database*, it provides a list of the average power consumption (Watt) of the most common appliances such as washing machine, oven, etc.;
3. Four worksheets, one for each type of *Distributed Renewable Energy (DRE) resource*, which allows to calculate the energy/gas production potential for a specific context, through the support of online databases and websites. The worksheets available are the following: worksheet for photovoltaic system sizing; worksheet for wind system sizing; worksheet for hydro system sizing; worksheet for biomass digester sizing.

The tool integrates databases and websites to get data on the local availability of renewable resources (e.g. Geographical Assessment of Solar Resource irradiation) (Figs. 7.7, 7.8, 7.9 and 7.10).

How to use the tool
First step is to define the energy load/need (worksheets 1) to determine the (potential) energy consumption of the system. To support the definition of the energy need in relation to appliances is possible to choose from the database of appliances (worksheet 2). After, it is possible to compare the energy load/need emerged, with the energy production potential of the DRE system designed (if any) to verify correspondence of energy need and energy availability. A second step is to size (or resize in case of existing) the DRE system according to the energy need to be satisfied. To do this, first step is to define the local renewable energy resource to be used: sun, wind, water and biomass (worksheets 3–4–5–6), and then to dimension the system according to the energy/load need. A final check is possible (worksheet 1) comparing the energy load/need and the energy production potential which has to (in average) correspond to the energy load/need.

Integrating the tool into the design process
The E.DRE tool is used in the *Design System Concept* stage to draft the new DRE systems, according to energy need and locally available resources.

Results
The result from the E.DRE tools is a preliminary sizing of new DRE systems, according to energy need and locally available resources.

Tool availability and required resources
The tool is available for a free download at www.lenses.polimi.it. It is available in digital version which could be used through a pc or a projector and requires internet connection to reach information from the online databases. The time required is approximately 60 min.

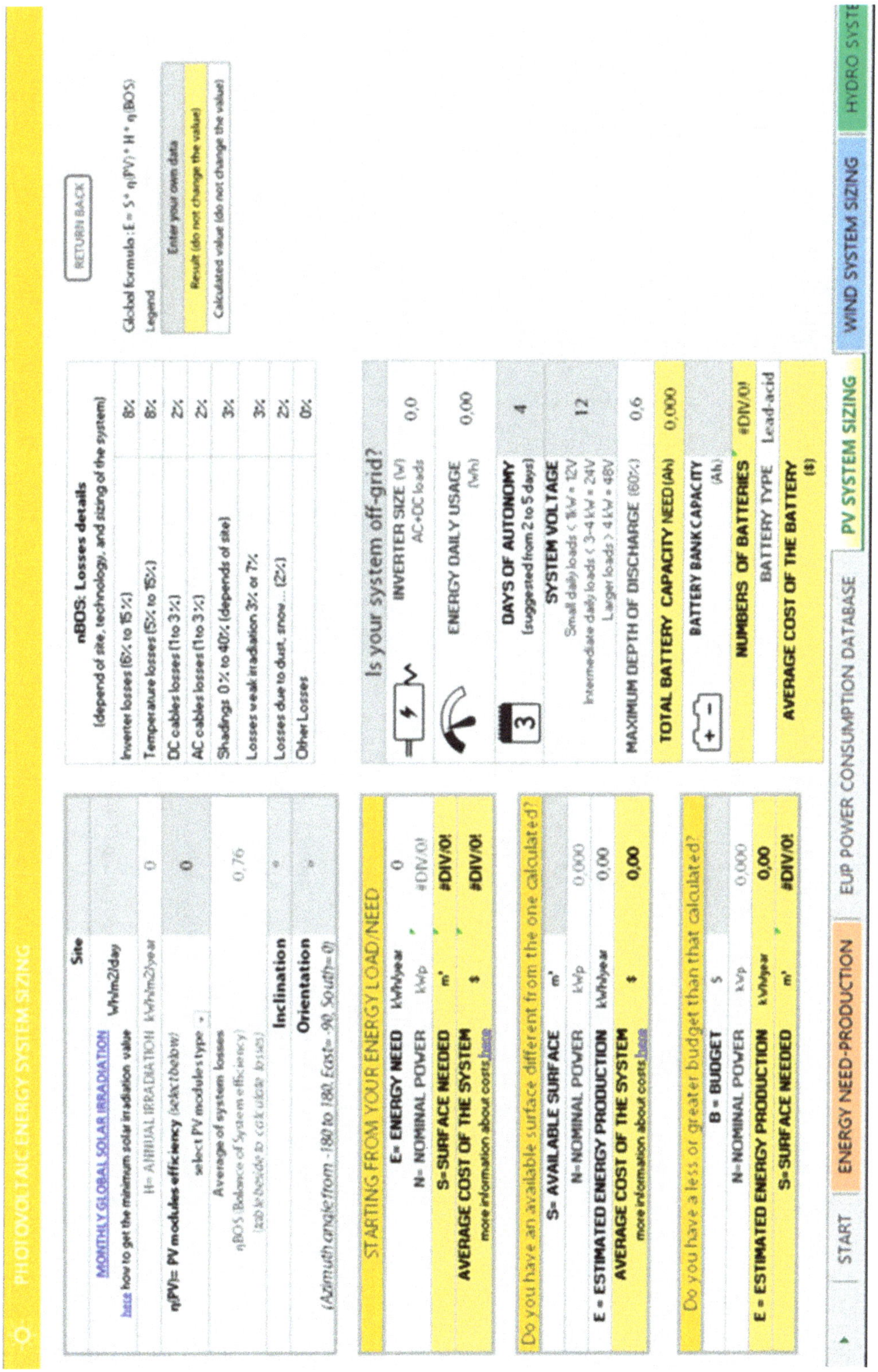

Fig. 7.7 Worksheet—solar energy. *Source* designed by the Authors

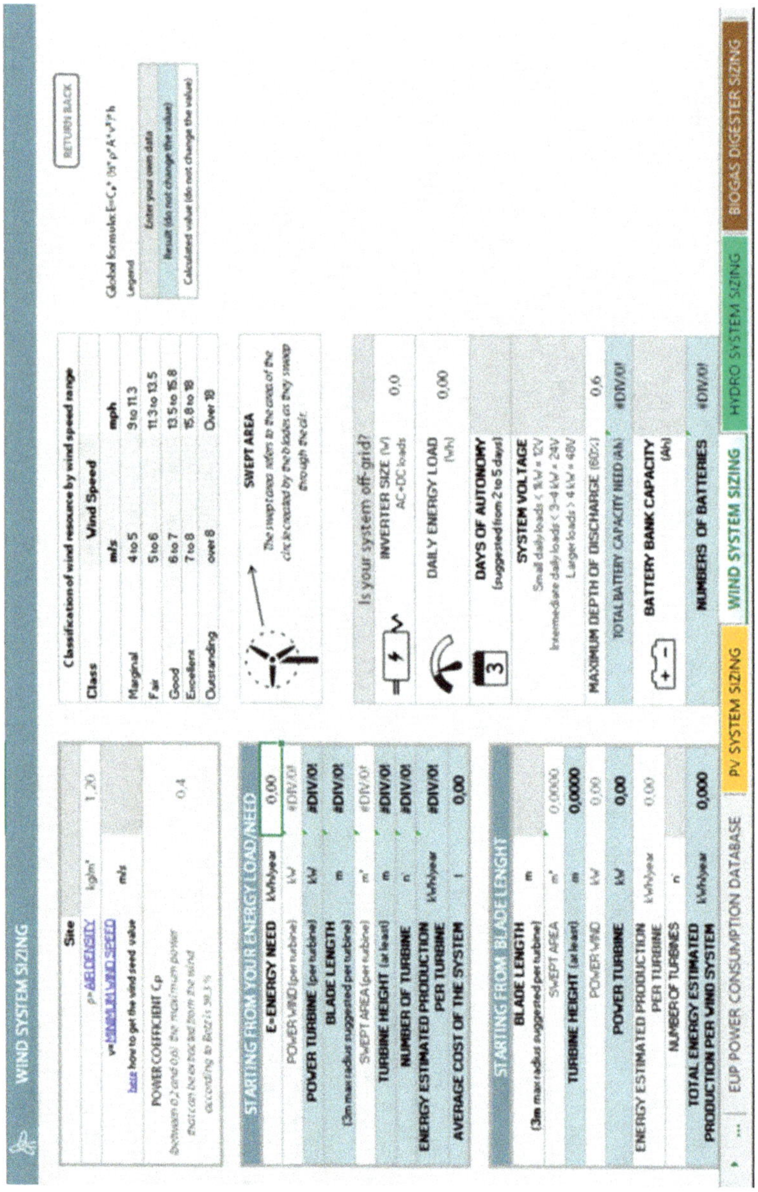

Fig. 7.8 Worksheet—wind energy. *Source* designed by the Authors

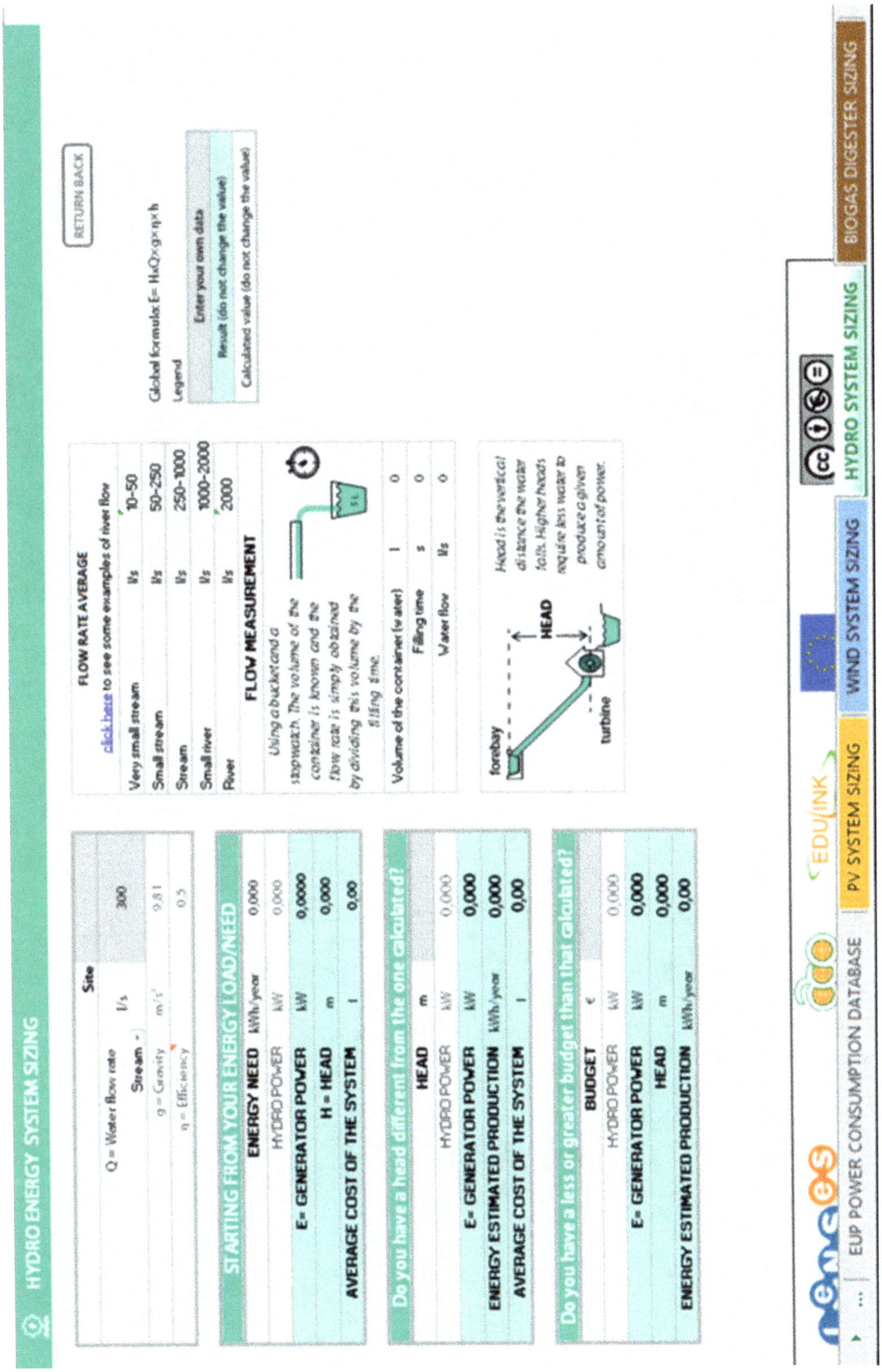

Fig. 7.9 Worksheet—hydro energy. *Source* designed by the Authors

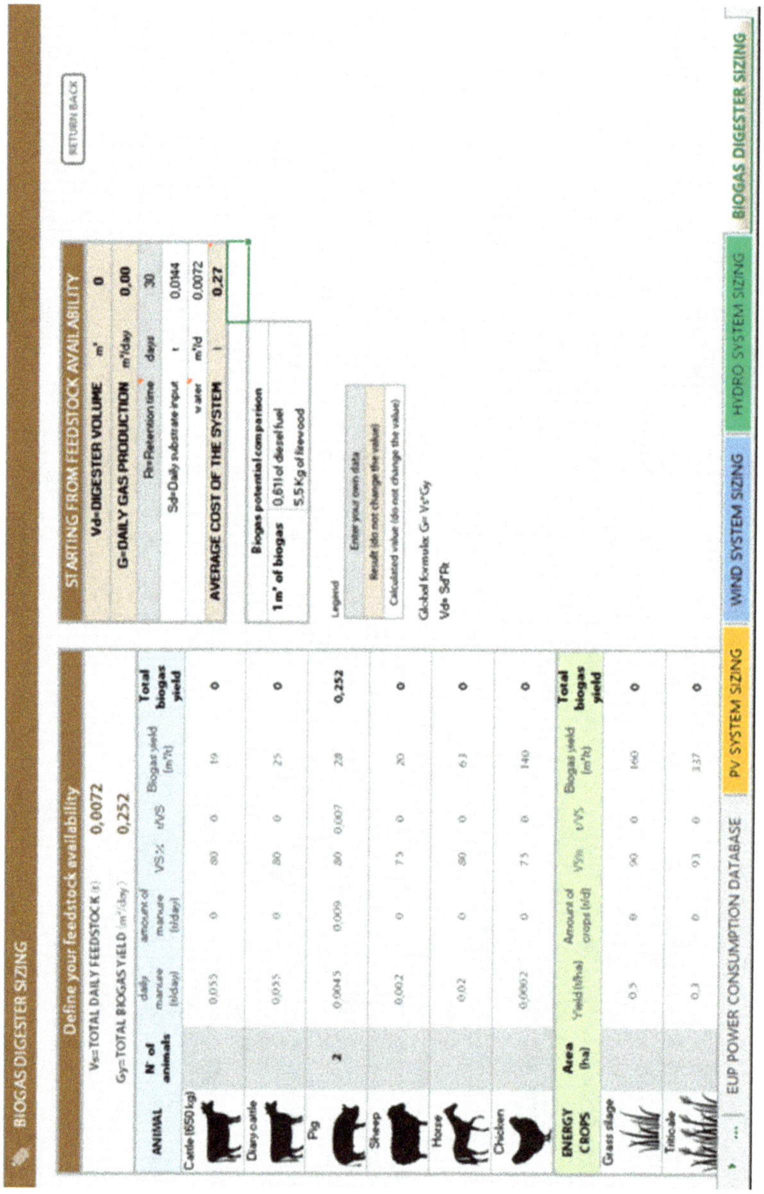

Fig. 7.10 Worksheet—biomass energy. *Source* designed by the Authors

7.2.5 PSS + DRE Innovation Map

Aims

The tool [5–7] can be used for classifying S.PSS models applied to DRE, positioning company's offers, analysing competitors in the market and exploring new opportunities. The tool can be also used for generating new concepts of S.PSS applied to DRE.

What it consists of

The tool is composed of the *Innovation Map*, the *Archetypal Models Cards*, *Stakeholder Cards* and a set of *Concept Cards*. The Innovation Map has been built as a classification system for S.PSS and DRE models [2] (see Sect. 4.4). The tool was built as a polarity diagram that combines different types of S.PSS models with the DRE energy systems and it can be used to position companies and new concepts according to the type of business model and the technology involved.

The vertical axis distinguishes the different types of S.PSS models, i.e. what is being offered to customers and what do they pay for. The different S.PSSs types on the Innovation Map help users to classify energy solutions based on what is the focus of the offer (product, use or result-oriented) and what is the payment structure (e.g. pay-to-purchase a product with financing services, pay-to-rent or pay-per-energy consumed). The vertical axis also encompasses ownership structure and environmental sustainability potential.

On the horizontal axis, the different types of DRE systems are illustrated: *mini kit*, *individual energy system*, *charging station*, *isolated mini-grid* and *connected mini-grid*. The horizontal axis encompasses also the type of target customers addressed in the S.PSS solution. It ranges from individual target (including the individual use of energy for households, entrepreneurs, productive activities, community buildings), to community target (which includes altogether a number of households, and/or productive activities, community buildings, public spaces, etc.) (Fig. 7.11).

The Archetypal Models Cards collect different types of S.PSS applied to DRE with corresponding case studies and a system map that illustrates how the system work (see Sect. 4.4) (Fig. 7.12).

The Stakeholders Cards aim at detailing actors and competitors involved in the energy scenario and at understanding their roles and responsibilities. This type of card can be used during the strategic analysis of competitors (see next section) (Fig. 7.13).

The concept Card aims at providing a template for generating new concept directions of S.PSS applied to DRE and it includes type of offer, network of providers, products, services, customers and payment channels. It can be used during the idea generation session (see next section) (Fig. 7.14).

How to use the tool

The tool can be flexibly used in different stages on the design process, from the strategic analysis (e.g. positioning company's offers and its competitors) to the idea generation and concept development phase.

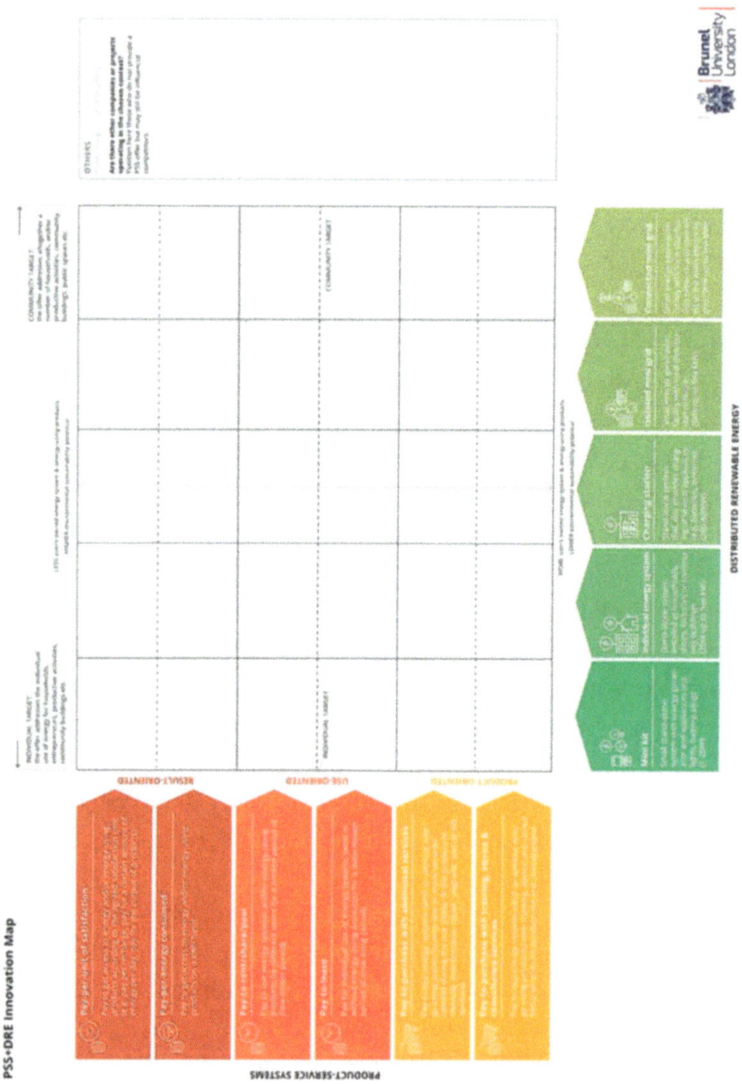

Fig. 7.11 PSS + DRE innovation map. *Source* Emili [7]

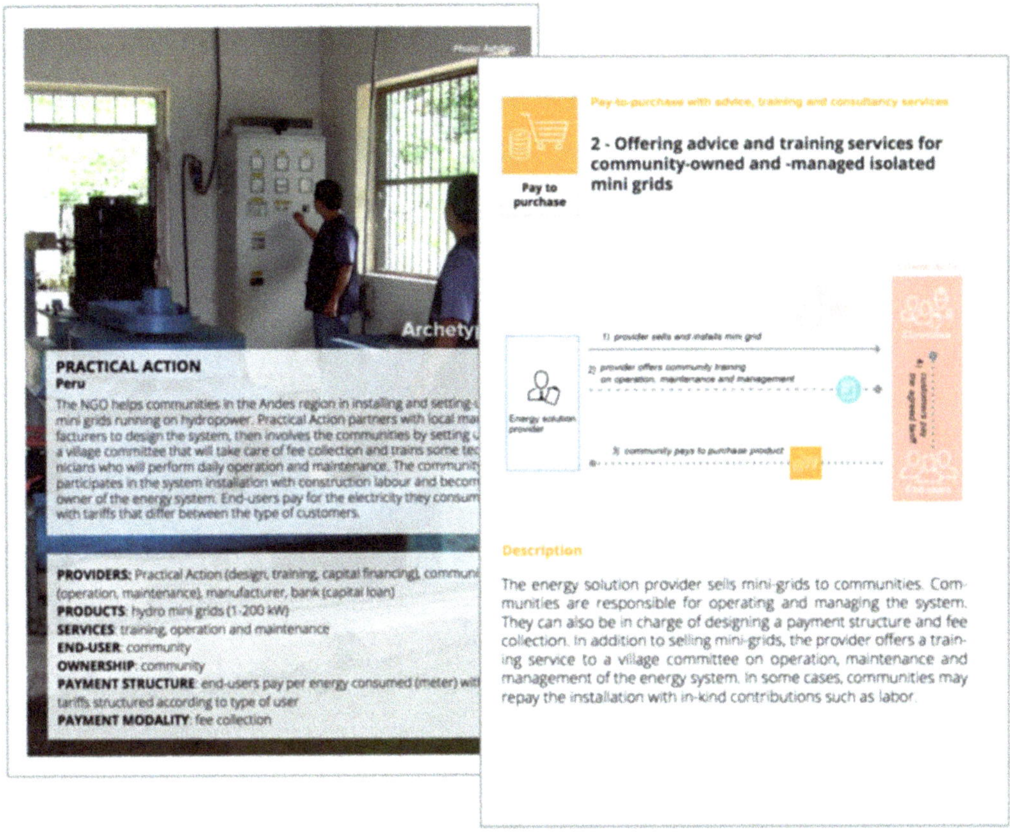

Fig. 7.12 Archetypes cards. *Source* [7]

Fig. 7.13 Stakeholders card. *Source* Emili [7]

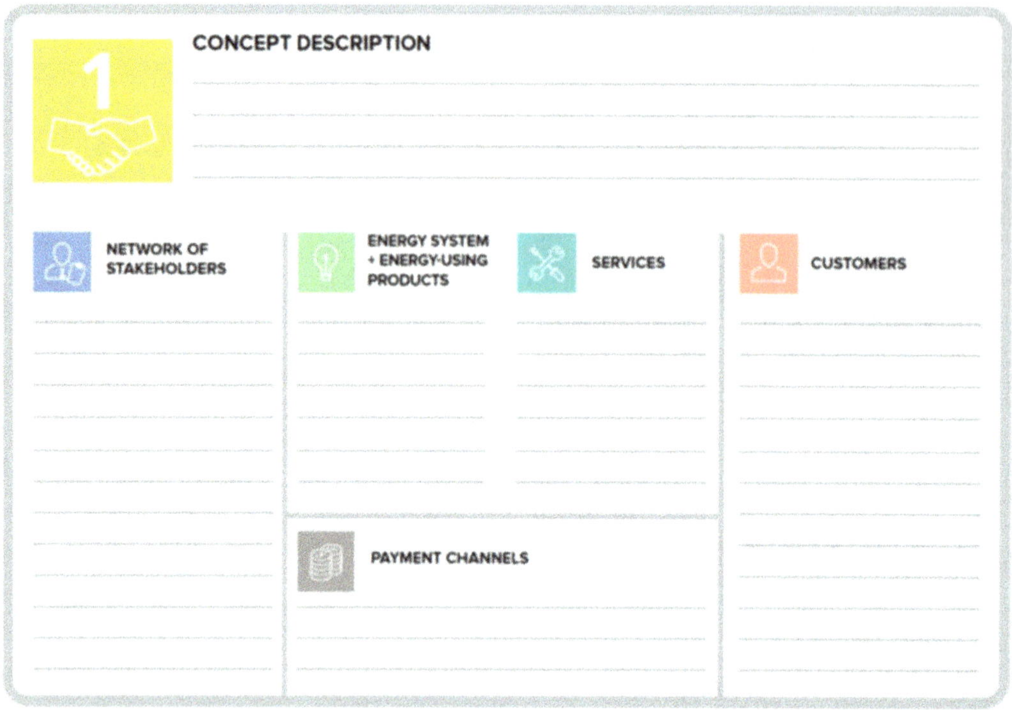

Fig. 7.14 Concept card. *Source* Emili [7]

Integrating the tool into the design process
The Innovation Map can be used for different purposes in the *Strategic analysis* and *Exploring opportunities* stages.

Strategic analysis
Position company's offerings on the map
The tool can be used to position a company's offerings according to the value proposition, type of energy system and target customer. Users can write down the company's offering on post-its (one offering per post-it), and place them on the map. The positioning should follow the type of S.PSS, i.e. product, use or result-oriented according to the specific payment structure and ownership model, and the type of DRE system involved in the solution. It should be highlighted that one company may have multiple offerings, and therefore these can be positioned on various parts of the Map (see Fig.3.1) (Fig. 7.15).

Map the competitors
Following the same criteria, companies operating in the selected context can be positioned on the Innovation Map, possibly using another colour of post-its. Users may want to focus on a specific technology (e.g. only mini-grid) or map all actors operating in a specific geographic area. If necessary, other offers that are not Product-Service Systems can be positioned in the box on the right-hand side of the Innovation Map (*Non-PSS offers*). These can include for example sale-based offers (e.g. sale of solar lanterns) or other complementary energy products (e.g. bioethanol fuel) (Fig. 7.16).

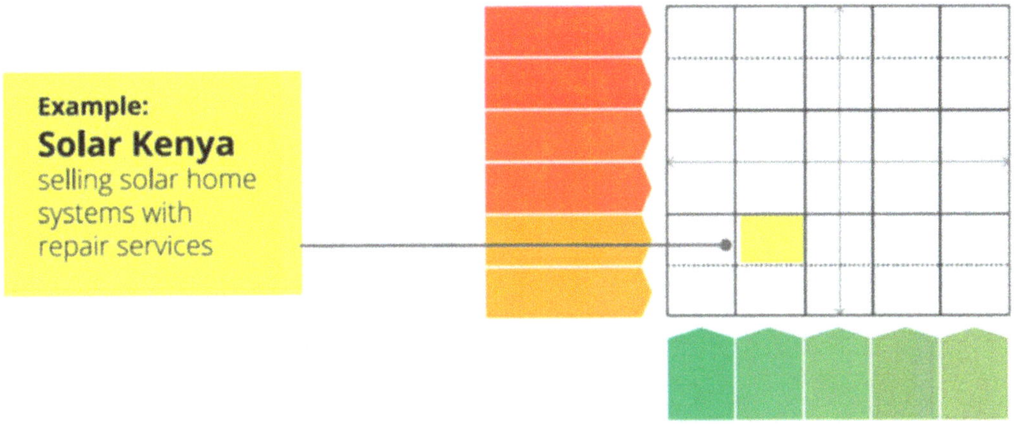

Fig. 7.15 Positioning of company's offerings on the innovation map. *Source* designed by the Authors

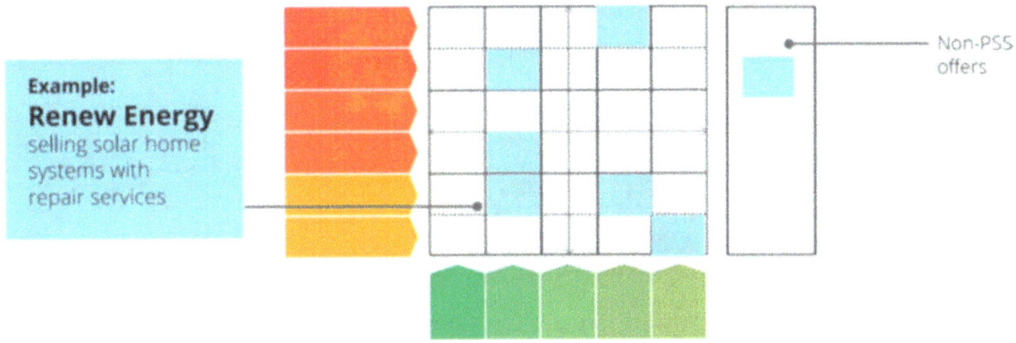

Fig. 7.16 Positioning of competitors on the innovation map. *Source* designed by the Authors

Strategic analysis of competitors: organisational form layer
To gather a deep understanding of the energy scenario, the tool can be used to detail the stakeholders that are providing energy solutions in a selected context and what roles and responsibilities they have. This phase aims at going more in-depth in analysing the target market by detailing the previously mapped solutions. The Stakeholder Cards can be used to define the actors involved and the roles they have. This phase can help users in understanding the main socio-economic actors operating in the energy sector in a specific area (Fig. 7.17).

Exploring opportunities
Select a promising area to explore
Having detailed the existing energy situation for the chosen context, users can focus on identifying promising areas to explore. This can be carried out but circling an area they want to focus on (Fig. 4.3). It could be a specific technology (e.g. individual energy systems) or a type of offering, or both. Areas that have not been explored by competitors in the same context may be a good starting point for tapping promising markets. It must be highlighted that the tool does not provide indications on how to identify promising areas. Instead, it acts as a framework to trigger and stimulate discussion among the design team (Fig. 7.18).

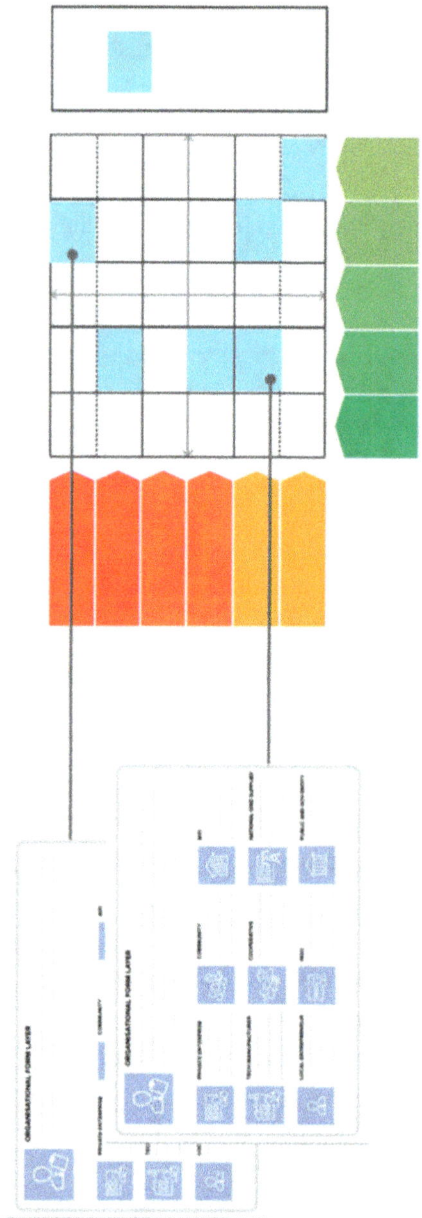

Fig. 7.17 Example of a completed stakeholder card. *Source* designed by the Authors

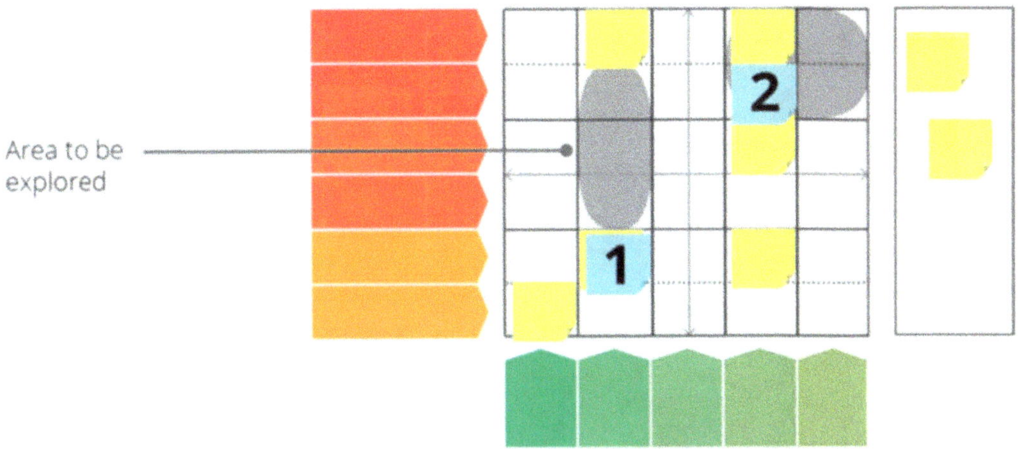

Fig. 7.18 Selection of promising areas to explore. *Source* designed by the Authors

Develop new concept directions

The Innovation Map can also support the design of new concepts of S.PSS applied to DRE. For this purpose, the Concept Cards can be used to write down ideas, starting by describing the general type of offer users intend to provide. Then, the corresponding number of the Concept Card can be positioned on the Map, following the same criteria used to map companies' offerings. At this stage, it is advised to generate several concepts, they will be selected and refined in the second moment.

Then, for each concept, the card should be filled out by writing down ideas on customers, products and services, stakeholders and payment modalities. At this stage, the aim is to consider the several elements of the design solution, without necessarily going into detail (Fig. 7.19).

Select the most promising concept(s)

Once the phases of strategic analysis and concept generation are completed, the Innovation Map should provide a visualisation of existing businesses/competitors, stakeholders involved, promising areas to explore and new business concepts. This

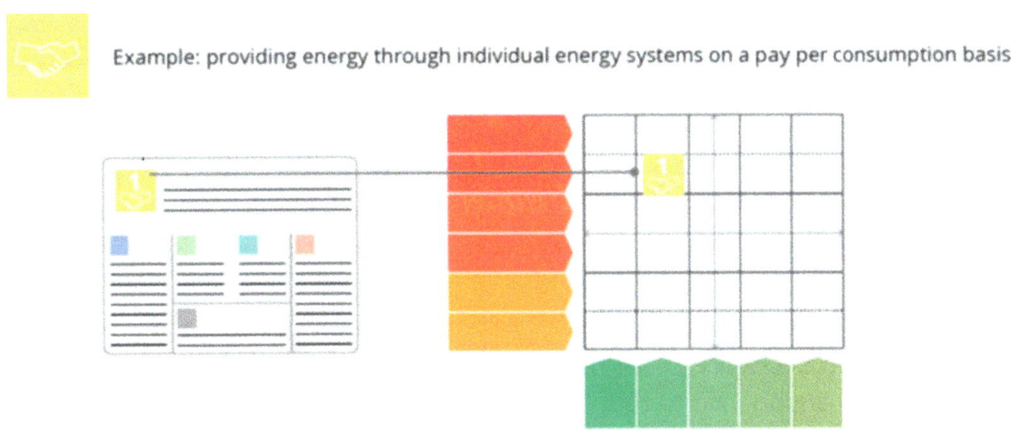

Fig. 7.19 Example of a completed concept card. *Source* designed by the Authors

Fig. 7.20 Example of a completed innovation map. *Source* designed by the Authors

can be the starting point for a discussion within the company's management team about which concepts are more promising, what influencing factors need to be considered and to eventually select one or more options for further detailing.

Results
At the end of the process, the Innovation Map provides a picture of the current situation (position of company's offerings, competitors and stakeholders involved) and a selection of promising areas to be explored. The Innovation Map also provides a first idea generation support to identify new business opportunities (Fig. 7.20).

Tool availability and required resources
The tool is available for a free download at www.lenses.polimi.it and on www.se4alldesigntoolkit.com. The tool has been designed to be used in workshops and (co)design sessions, therefore, it is preferable to print it in a large format (at least A1). The time required for using the Innovation Map can vary, but a minimum of 2 h is suggested to complete all design phases.

7.2.6 S.PSS + DRE Design Framework & Cards

Aims
The tool [3, 7, 8] can be used to support the generation of ideas on specific aspects of S.PSS applied to DRE (network of providers, customer, products and services, offer and payment channel), and to bring an initial concept idea to a detailed concept.

What it consists of
The tool is composed by a Design Framework, a set of Cards and a Design Canvas (Fig. 7.21).

The Design Framework
The Design Framework visualises the main elements characterising S.PSS applied DRE models, which are organised in six 'building blocks'. Each building block includes specific elements to be considered in the design, as described below.

Network of providers It refers to the actors involved in providing the energy solutions and it includes private enterprise, technology manufacturer, community, local entrepreneur, Non-Governmental Organisation (NGO), Cooperative, Micro-Finance Institution (MFI), public and governmental entity and national grid supplier.

Products It refers to the combination of energy system/s (including renewable energy sources) and energy-using product/s. **Energy systems** include stand-alone systems (mini kit, individual energy system, charging station) and grid-based systems (isolated and connected mini-grid). Energy systems also included the types of **renewable energy sources** used for DRE: solar, hydropower, biomass, wind or hybrid sources (i.e. combination of different renewables). **Energy-using products** refer to the appliances that can be included in the offer in combination with the energy systems (i.e. generator). These might include lantern, lights and bulbs, battery, phone charger, radio, TV, fan, IT and computer devices, etc.

Services The service category includes consultancy services (training, financing) and services provided during or at the end of the product life cycle (installation, maintenance and repair, product upgrade, end-of-life services).

Offer This building block refers to the different types of S.PSS offer that can be applied to DRE models. Their classification is divided into product-oriented (pay-to-purchase with training, advice and consultancy services; pay-to-purchase with additional services), use-oriented (pay-to-lease; pay-to-rent/share/pool) and result-oriented S.PSSs (pay-per-energy consumed; pay-per-unit of satisfaction).

Customers It refers to the type of target customers addressed by the S.PSS solution and includes individual household, productive activity, local entrepreneur, public buildings, community, public and governmental entity, mix of target customers.

Payment channels This building block refers to the different ways customers pay for the energy solution. It includes cash, credit, mobile payments, scratch cards and energy credit codes, in-kind contribution, fee collection and remote monitoring as an activity supporting payment.

For each building block, the Framework provides a series of questions that should guide the user in the design process. For example, the network of providers building block presents the following questions: 'Who are the actors involved in the provision of the energy solution? What are their roles and responsibilities? What partnerships can be established?' (Fig. 7.22).

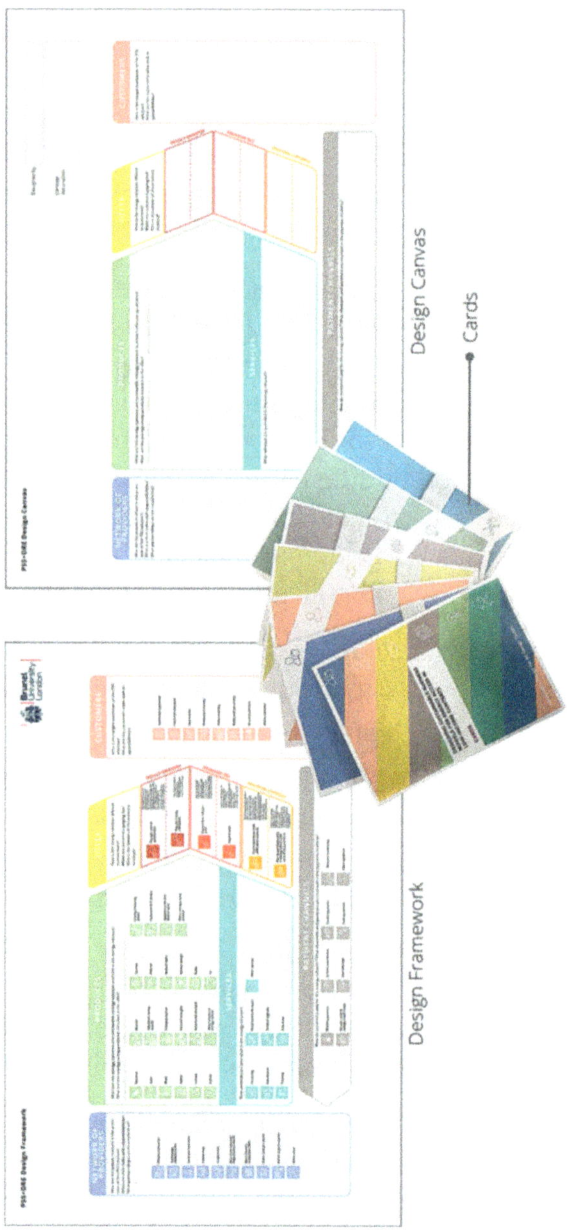

Fig. 7.21 The design framework, design canvas and cards. *Source* designed by the Authors

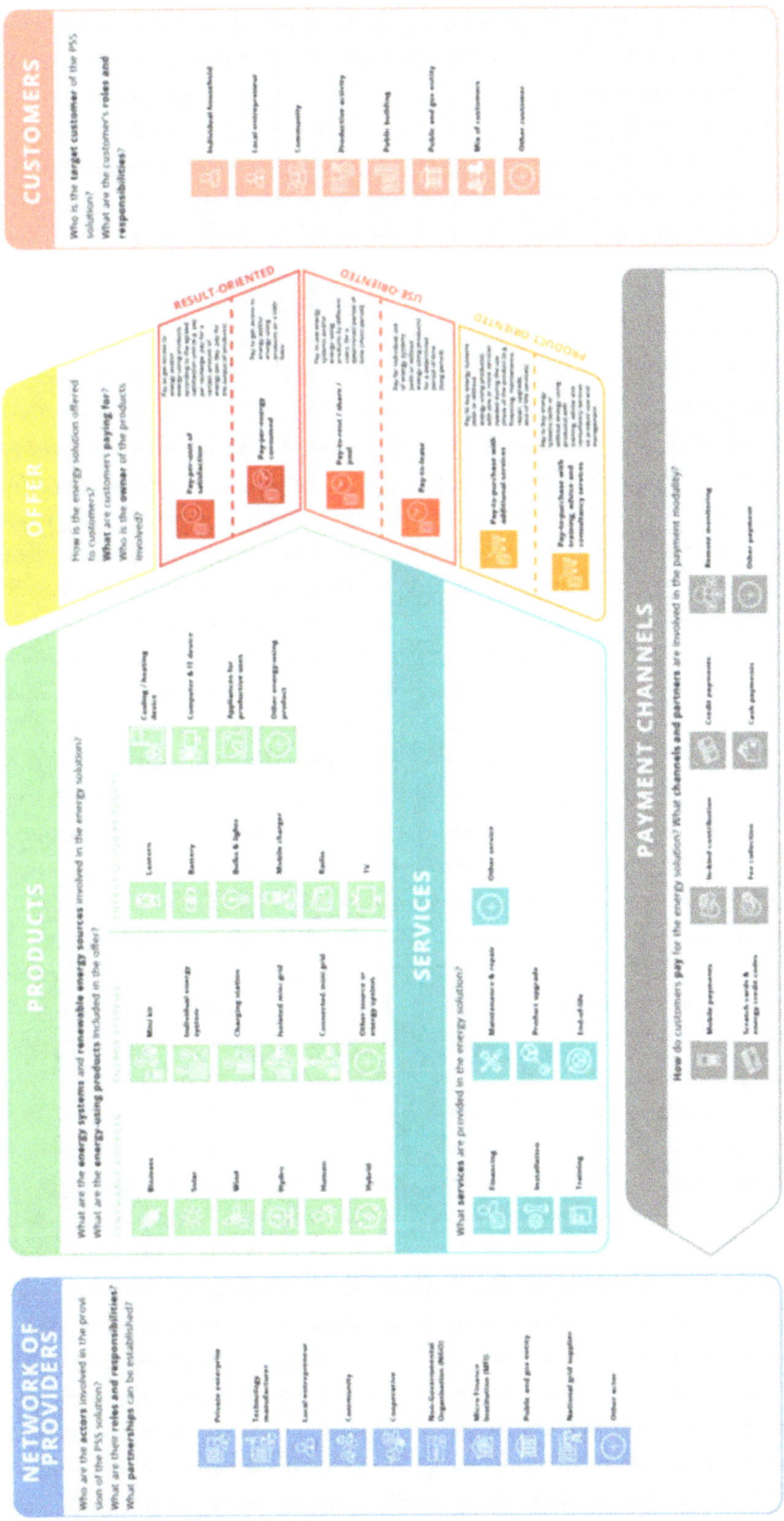

Fig. 7.22 Design framework. *Source* Emili [7]

Cards

The Cards have been developed with the aim of providing support to companies and practitioners in designing the S.PSSs applied to DRE. In particular, they collect critical factors, guidelines and successful examples of S.PSS applied to DRE in low-income and middle-income contexts. In particular, the Cards summarise and organise in a clear and meaningful way the existing knowledge developed on S.PSS applied to DRE (i.e. critical factors and case studies [3], see Sect. 4.5), so that it can be used to trigger the generation of ideas. Cards are organised according to each building block (Fig. 7.23).

Each group of cards is provided with an Intro Card that specifies what information you can find in there and how to use it. A general structure of the elements contained in the cards is illustrated below (Fig. 7.24).

Design Canvas

The Design Canvas is an empty Framework that should be used in the concept generation phase to position post-its and write down ideas. The Canvas follows the same structure as the Design Framework and distinguishes S.PSS + DRE building blocks—network of providers, products, services, offer, customers, payment channels. It is also provided with some questions to guide the design process (Fig. 7.25).

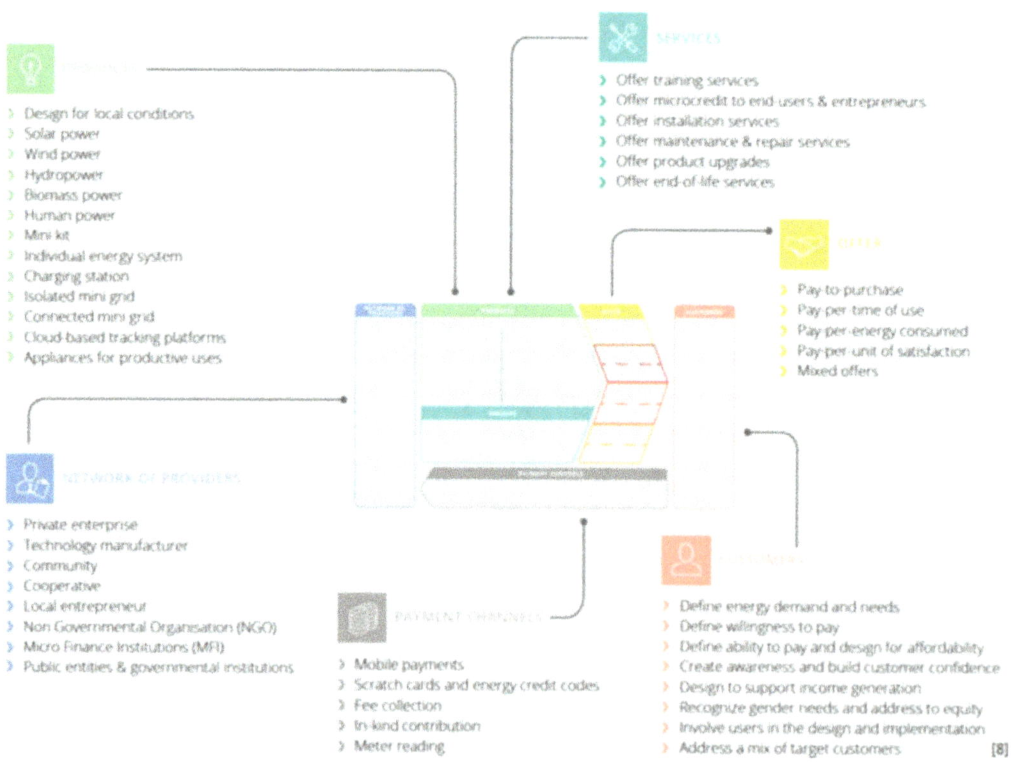

Fig. 7.23 List of cards for each design element. *Source* Emili [7]

Fig. 7.24 Example of cards' structure. *Source* Emili [7]

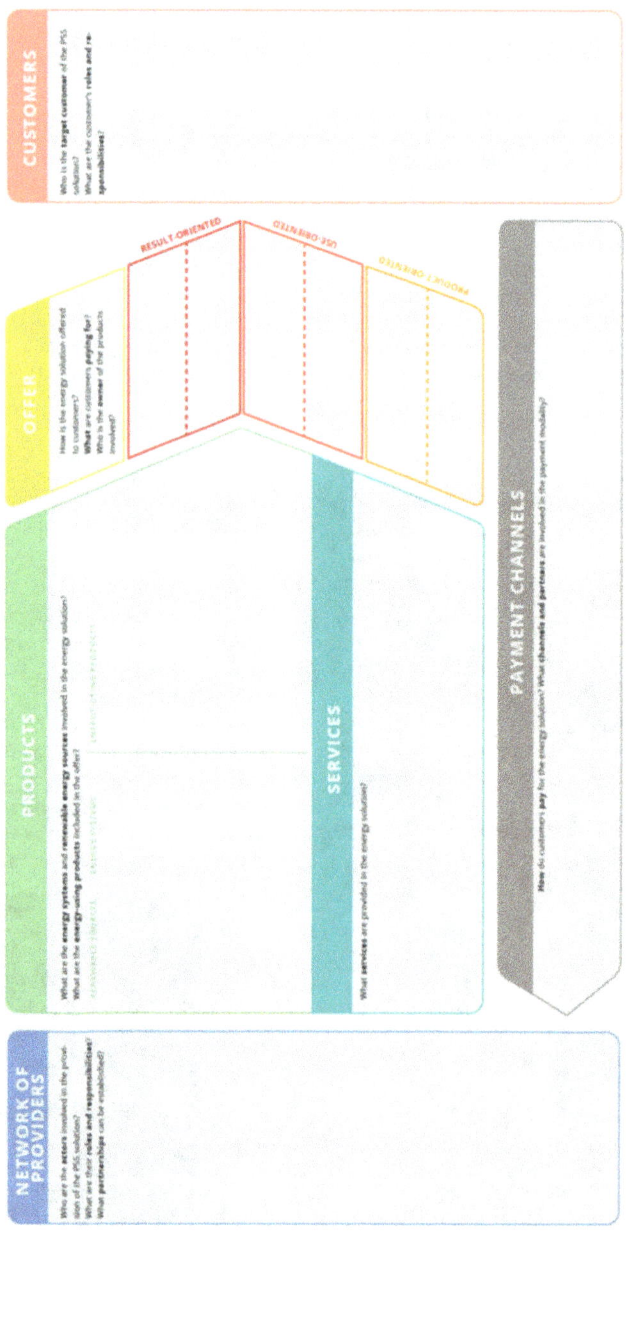

Fig. 7.25 The design canvas. *Source* Emili [7]

How to use the tool

The Design Framework and Cards has been developed to be flexibly applied according to users' needs. In particular, the tool finds application for:

- *start-up a new business*: to support the design and detailing of new business models from scratch.
- *refine and reorient existing solutions*: the tool can be used to focus on specific aspects of an existing business model. For example, a company might already have an offer in place but may want to improve aspects related to the payment channel.

This section illustrates how the Design Framework and Cards can be integrated in the SD4SEA design process and what outcomes can arise from its application.

Integrating the tool into the design process

Exploring opportunities

Generate ideas

The tool can be used in the beginning of the Exploring opportunities stage to support brainstorming sessions to generate ideas on the various building blocks of the Design Framework. In other words, the tool can be used when there is not any agreed concept direction, to inspire idea generation looking at the various aspects of S.PSS applied to DRE. Ideas then can be reviewed, selected and combined to develop initial concept directions. The idea generation process does not have to follow a specific order; it is possible to start from any building block.

System Concept Design

Detail initial concepts

The main application of the Design Framework and Cards is the detailing of an initial concept idea. In fact, the tool allows to go in-depth in all the building blocks and to generate ideas for each of them. This activity can be carried out after having used the Innovation Map to generate a concept idea, or if the designer/s has already a draft idea of the business model they would like to detail. After the idea generation, ideas are reviewed to select the most promising ones to be integrated into a detailed concept design.

Improve specific aspects of an existing solution

The tool can also be applied to brainstorm on a specific aspect of an existing S.PSS solution. For example, a company already delivering a S.PSS solution may want to improve the payment modality, and they can use the tool focusing only on the Payment Channels building block to get inspired by the guidelines, case studies and suggestions (Fig. 7.26).

The use of the tool does not require following a specific order for the idea generation. Users are encouraged to decide the starting point they prefer. The design process can be, therefore, carried out in an unstructured way, for example, browsing Cards and using the Framework as a reference, and then writing down ideas on post-its, positioning them on the Canvas (Fig. 4.11).

Fig. 7.26 Positioning ideas on the design canvas. *Source* designed by the Authors

Results
At the end of the design process, all elements of S.PSS applied to DRE should be detailed with selected ideas (among the ones generated in the activity), and the questions provided on the Canvas should be answered. The tool can be used in combination with other tools and resources; in fact, concepts generated with the tool might require further evaluation in terms of financial sustainability, technical feasibility, presence of appropriate regulations and other external factors.

Tool availability and required resources
The tool is available for a free download at www.lenses.polimi.it and on www.se4alldesigntoolkit.com. The tool has been designed to be used in workshops and (co)design sessions, therefore a printed format is preferable: the Design Framework should be at least A2, the Design Canvas can be printed in A1 and the Cards can be printed on A4 and folded.

To use the Design Framework and Cards, we suggest from a minimum of 2 h to grasp the most essential aspects; to a 8 h to go in-depth and detail every building blocks. We also suggest that the idea generation is carried out in multidisciplinary teams to maximise innovation potentials.

7.2.7 The Energy System Map

Aims
The Stakeholder System Map tool, developed by [7, 9] to visualise the network of stakeholders in a S.PSS solution, and their interactions (in terms of flows of goods, materials, services, money, work and information), has been adapted to be specifically used for S.PSS applied to DRE The Energy System Map [4] is presented as a visualisation tool, with its specific set of icons, flows and rules that aims at supporting (co)designing and visualisation of S.PSS applied to DRE models.

It is, therefore, a support tool for

- *Designing* because representation is a means of structuring thought and facilitating the resolution of problems;
- *Co-designing* because a standard language is used, which can, therefore, be shared by all the design team members or the different actors involved, supporting the strategic conversation among them;
- *Communicating* because it enables unambiguous visualisation of the designed solution (as well as its evolution).

What it consists of
The tool allows the development of a graphic representation showing

- the socio-economic actors involved in a S.PSS solution (both primary and secondary stakeholders);
- and the various interactions among these actors, in terms of flows of goods, materials, services, money, work and information.

The tool is a representational tool that can be described as both codified and progressive. It is a codified system in the sense that it can be considered a 'technical drawing' representing the actors involved in a S.PSS in a standardised and comparable way. It is progressive in the sense that it is a 'formalisation-in-progress' of the solution actor map giving an increasingly accurate picture of the project as it develops.

The tool is composed by a set of icons (to represent socio-economic actors as well as the various physical and intangible elements of the S.PSS), arrows (to represent the various types of flows/interactions between the actors), a template to be used in the design process and a set of rules for the visualisation and a set of rules to visualise them. Icon is characterised by colour-coding and a short text describing the actor, product or activity (Figs. 7.27 and 7.28)

How to use the tool

The tool requires the use of a slideshow software (e.g. Microsoft PowerPoint, or the equivalent in Open Office), but a printed version can also be used. The tool is based on some specific rules to be followed (Fig. 4.12) that aim at standardising each S. PSS + DRE model.

Each actor is represented by one icon, made of three elements.

- The structure, which indicates the typology of actor, e.g. private enterprise, public institution, community, etc.
- The colour, which defines the type of icon, i.e. services (light blue), products (green), etc.
- The slogan, which specifies the actor activity: energy solution provider, micro-finance institution, etc (Fig. 7.29).
- The energy solution provider/s, which can include a single actor or a partnership of actors, is represented on the left-hand side of the map and it is characterised by a violet colour;
- The customer is always placed on the right-hand side of the map and it is characterised by a pink colour;
- Ownership of the energy system and energy-using products are described with corresponding colours;
- Flows of products and services are pictured in the top-middle part of the map, showing transactions between provider and customer—Payments are described in the bottom of the map, showing what the customers pay for and what modalities/channels are used;
- In order to facilitate the reading of the map, flows are ordered with progression numbers.

The nature of the flows between the different actors is marked by different arrows (Fig. 7.24):

- The full, thick arrow indicates material flows (components, products, etc.);
- The fine, square-dotted arrow indicates information flows;
- The fine, round-dotted arrow indicates money flows;
- The full, thick arrow with a diamond at its tip indicates workflows (Fig. 7.30).

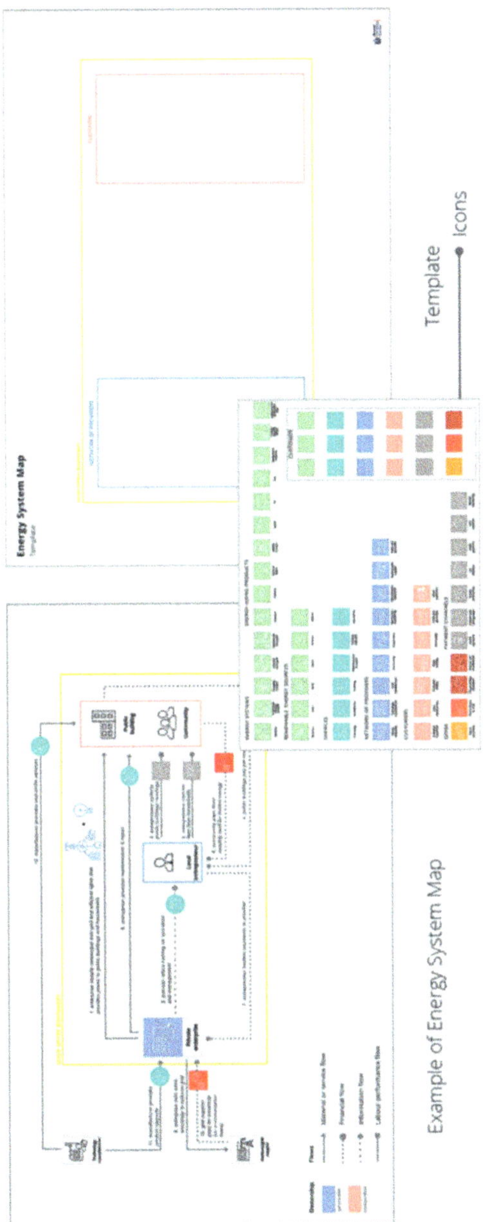

Fig. 7.27 The energy system map tool: icons, example and template. *Source* designed by the Authors

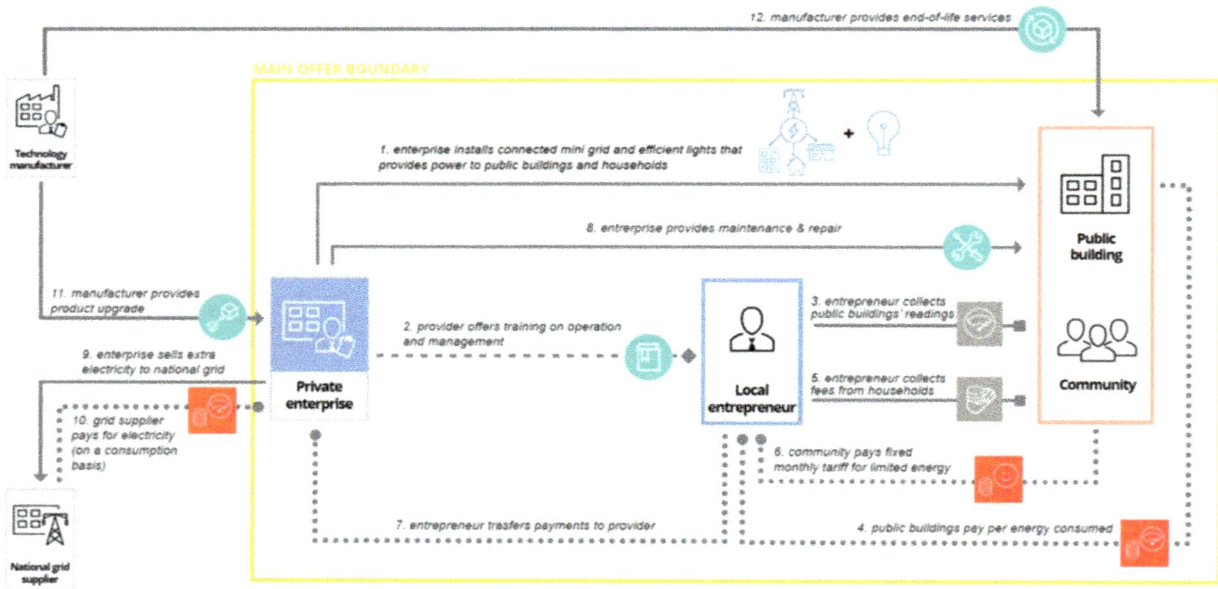

Fig. 7.28 Example of energy system map. *Source* designed by the Authors

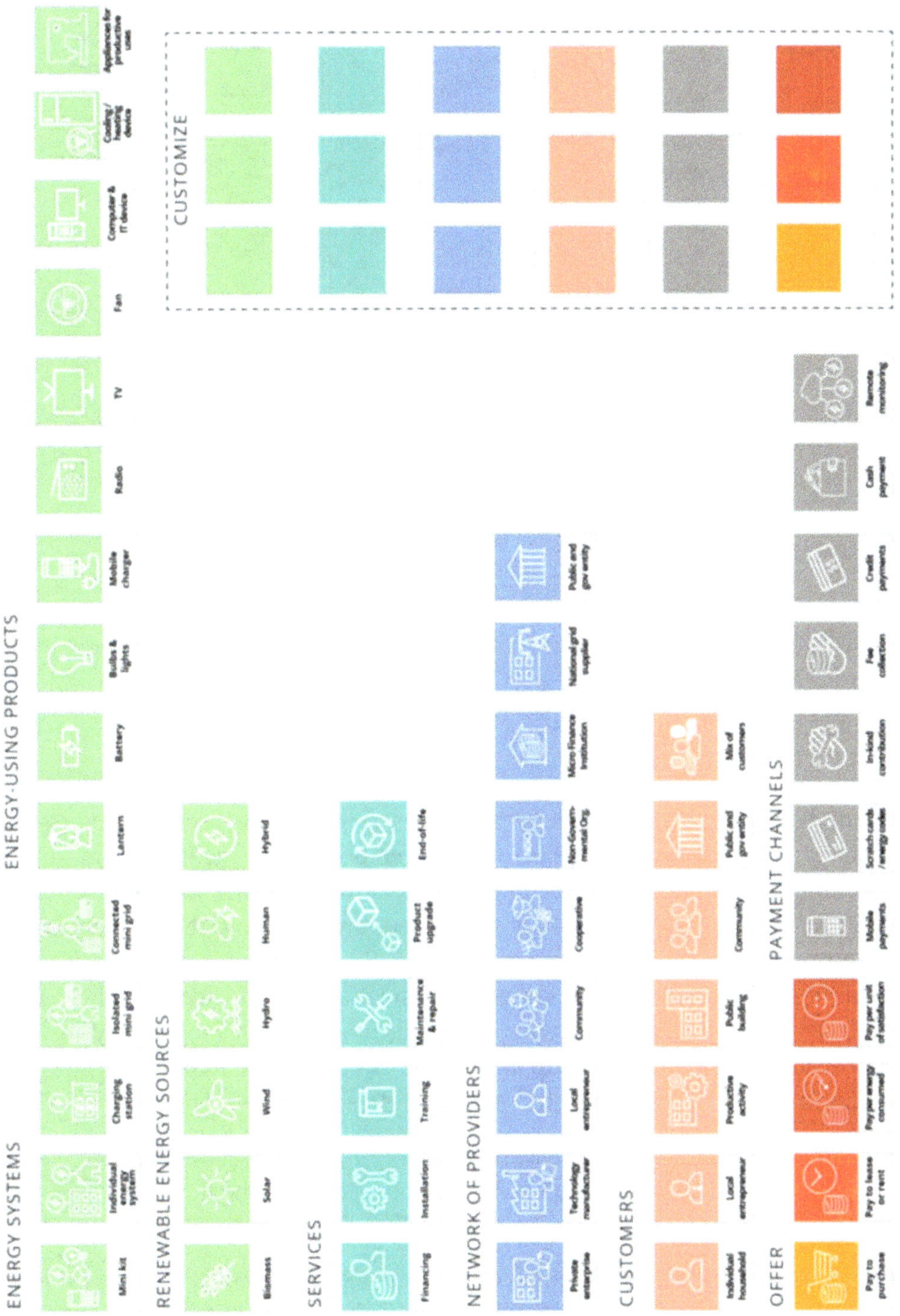

Fig. 7.29 Icons. *Source* Emili [7]

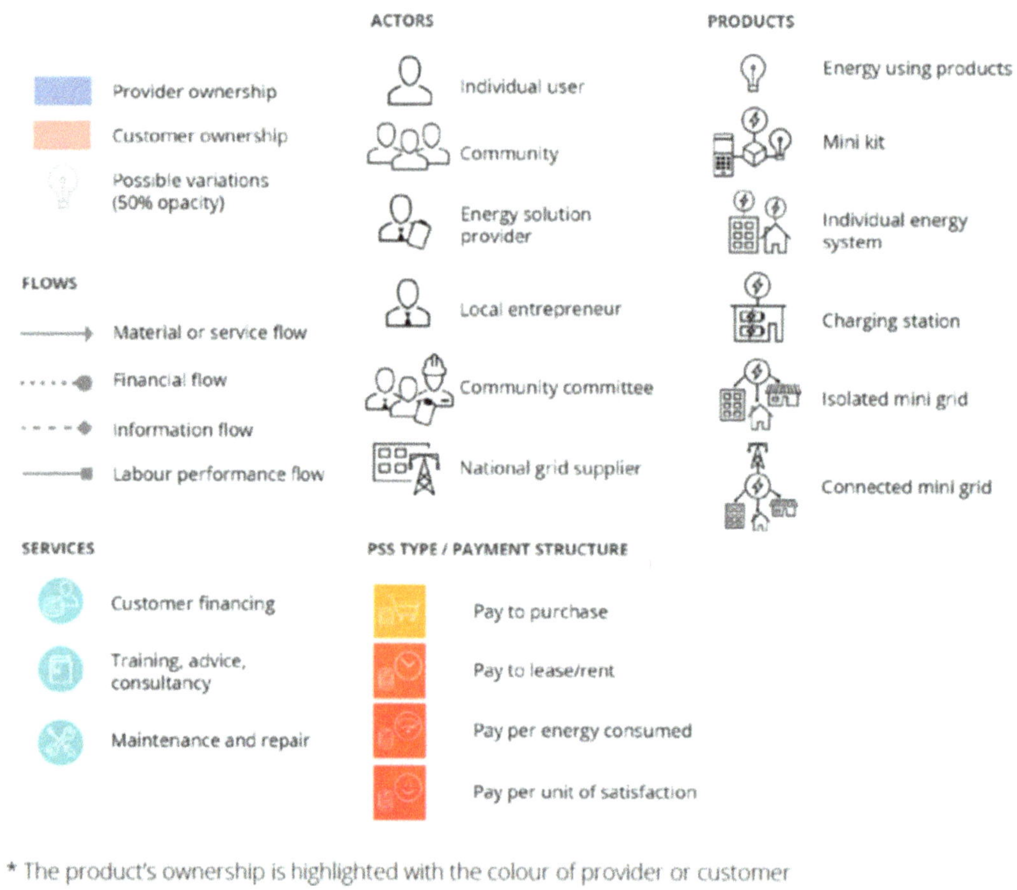

Fig. 7.30 Legend for the energy system map. *Source* Emili [7]

- System boundary by convention, the limit of the slide or the sheet is the boundary of the system, while a 'main offer boundary' includes core actors performing the system. Main actors, their relationships and the main offer to customers are represented within a defined area (yellow box). Secondary stakeholders and their involvement in the S.PSS solution can be positioned outside this area, usually represented with smaller icons to indicate their subordination. This would include, for example, financing and regulatory institutions which are involved in supporting the S.PSS solution but they are not directly involved in providing the offer to end-users (Fig. 7.31).

Integrating the tool into the SD4SEA designing process

The Energy System Map can be used at various stages of the designing process. In the **Strategic analysis,** it can be used to describe

- The current stakeholder value chain of the organisation(s) involved in a project;
- The stakeholder value chain of S.PSSs provided by competitors or of cases of excellence.

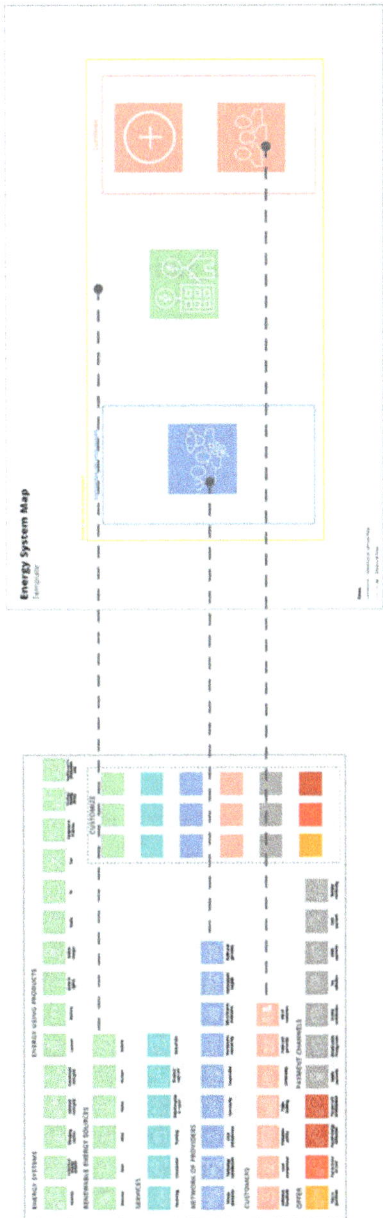

Fig. 7.31 Positioning icons on the template. *Source* designed by the Authors

In the **System concept design,** it can be used to

- Formalise the initial S.PSS ideas emerging, by visualising the key stakeholders involved in the solution;
- Detail the initial ideas emerging, identifying the main and secondary actors and their interaction flows.

In the **Design System details,** it can be used to

- Further detail the configuration of the system, by visualising all the actors involved and their interactions.

Results
The result is a map that shows the various socioeconomic actors that take part of the system and their interactions (in terms of material, information, money and workflows). This map becomes more and more detailed as the project evolves.

Tool availability and required resources
The tool is available for a free download at www.lenses.polimi.it and on www.se4alldesigntoolkit.com. The tool has been designed to be used in workshops and (co)design sessions, therefore, a printed format is preferable. Alternatively, the tool can be used in its software version (Microsoft PowerPoint), which allows users to modify icons and personalise their Energy System Map. The time required to generate a *System Map* is approximately 30 min. For more complex systems additional time may be required.

7.2.8 Innovation Diagram for S.PSS&DRE

Aims
The Innovation Diagram for S.PSS&DRE [13], is a tool to analyse competitor's energy solutions; as well as to orient the design of new S.PSS applied to DRE concept. The tool allows selection and clustering of (environmentally, socioethically, energy) sustainable ideas within polarity diagram, and starting the design of new S.PSS applied to DRE concepts. Furthermore, it provides the characterization of the designed S.PSS applied to DRE concepts through a set of labels and suggestions. The Innovation Diagram for S.PSS&DRE is presented as a co-design tool favouring a deep understanding of the solution/concept while facilitating collaborative processes and discussions among stakeholders (Figs. 7.32 and 7.33).

What it consists of
The tool is composed by three worksheets for existing energy solutions, for competitors' energy solutions, for new concepts. Each worksheet is based on the following structure: title + proposer + unit of satisfaction + polarity diagram + profile (with labels) + short description. The worksheet for new concepts includes post-it to stick new ideas from the Sustainable Energy for All Idea Tables tool. Two additional worksheets with labels and instructions are available to fill the profile section of the tool.

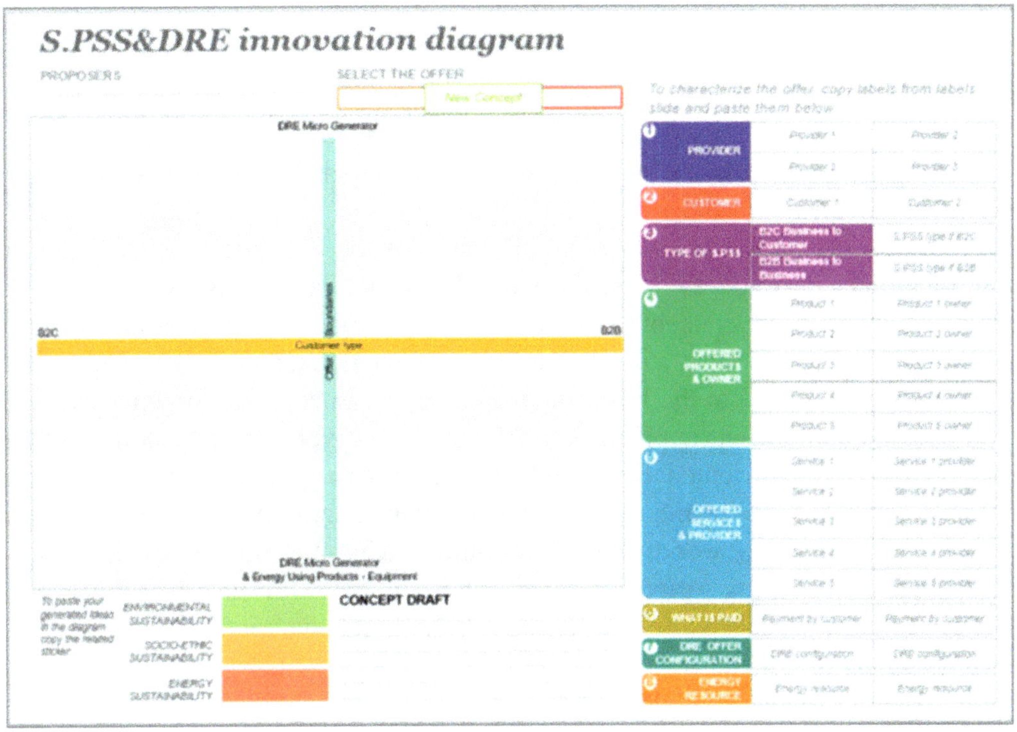

Fig. 7.32 Innovation diagram for S.PSS&DRE. *Source* designed by the Authors

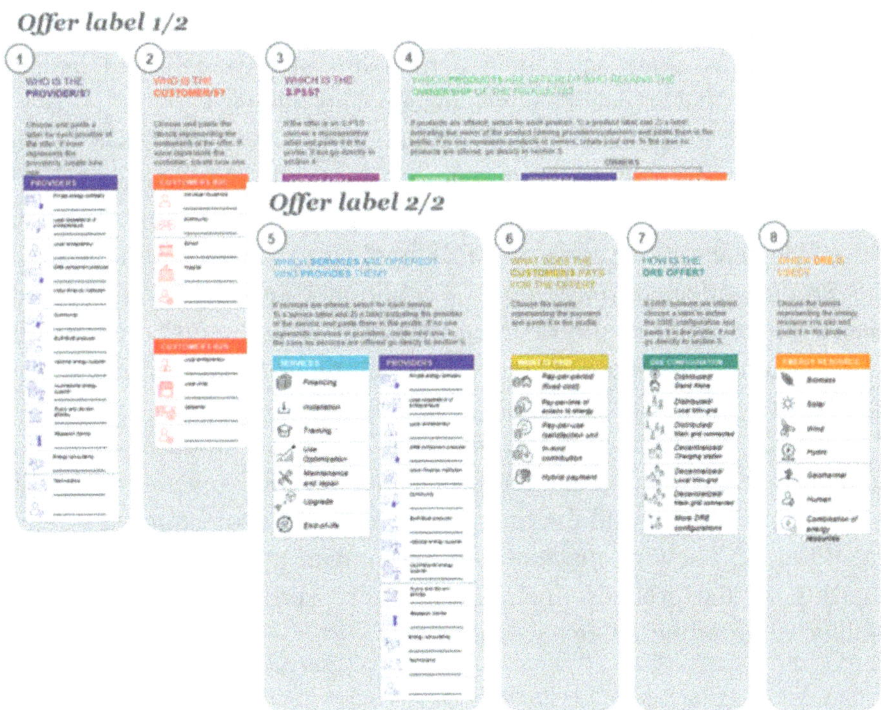

Fig. 7.33 Labels to support the innovation diagram for S.PSS and DRE. *Source* designed by the Authors

In each of the worksheet the following could be found:

Title depending on the worksheet the title is the name of the solution/concept that will be explored.

Proposer means the name/s of who is using the tool.

Unit of satisfaction is the need satisfied/to be satisfied (e.g. access to energy, have clean clothes).

Polarity diagram the polarity diagram (same of the Scenario one) is based on four quadrants built on two axes (a) the horizontal axis defines to whom is addressed the solution/concept end user (B2C), or small entrepreneur/small business (B2B) (b) the vertical axis defines how much is extended (boundaries) the solution/offer is related to the Distributed Renewable Energy micro-generator (e.g. solar panel system plus its appliances such as storage, inverter, wires, etc.), or to the sum of both the Distributed Renewable Energy micro-generator and the related Energy-Using Products or Energy-Using Equipment (e.g. phone and television are Energy-Using Products; woodworking machine, sewing machine are Energy-Using Equipment). Due to the variety of actors who can deal with energy solutions, is relevant to consider that actors can play in the polarity diagram even though they are not directly offering Distributed Renewable Energy micro-generator, and neither Energy-Using Products or Equipment. For example, a consultancy on energy services could be positioned on one pole or the other on the typology of energy services.

Profile (with labels) the profile presents a table with empty spaces to be filled with the following *key information* regarding the energy offer[3]:

- *Provider/s* it refers to the providers involved in delivering the energy solution and could be one as alone actor or a partnership of providers and includes energy companies, NGOs, energy consultancies and others;
- *Customer/s* it refers to the customer of the energy solution and can be a final customer (B2C) as a household, a community, a school, and so on; or a small entrepreneur/small business (B2B) such as a company, a local shop and others;
- *Type of S.PSS* it refers to the type of S.PSS applied to the energy solution and could be Product-oriented (pay-to-own + additional services as installation, maintenance, repair, and others) or Use-oriented (pay-to-lease/share + training to install, maintain, manage and so on) or Result-oriented (pay-per-use to reach a specific final result/satisfaction unit);
- *Offered product/s (and related ownership)* it refers to products which are included in the energy solution and integrates both the DRE generator (e.g. solar panel + wires and storage) and the Energy-Using Products or Equipment (e.g. a phone is a Product; a sewing machine is an Equipment). It is required to define the ownership of the products included to lately verify the innovative and sustainability value of the energy solution;

[3]*To increase readability of this section, we will use the term 'energy offer' both to refer new concepts or to existing energy offers, or competitor's energy offers.*

- *Offered service/s (and related provider)* it refers to services which are provided by the energy solution such as financial services, training services, maintenance services and so on. It is required to define who among the stakeholders is delivering the service to define and verify the role of each stakeholder in the energy solution;
- *What is paid* it refers to what the customer (B2C–B2B) pays to access the energy solution as pay-per-period, pay-per-use, pay-per-time, in-kind payment, payment with financial support, or a mix of different payments;
- *DRE system configuration* it refers to how the energy solution is structured. The options include distributed stand-alone system (e.g. solar home system, solar lantern), decentralised stand-alone system (e.g. energy recharging centre), distributed-decentralised systems connected through mini-grid as well as distributed-decentralised systems connected to the main-grid;
- *DRE source* it refers to the energy source used to power the energy solution as solar, wind, hydropower, biomass, and others and could be a single source or an integrated mix of them.

Labels and instructions
The labels are divided as per the profile key information (see above) and offers for each of them a series of variable solutions, e.g. for the customer there are several labels such as, community, household, etc., the same is for all key information. To facilitate the use of the label a question and guideline for each key information is provided.

Short description
The short description is no more than 200 characters, to be used to present the solution/concept highlighting the main innovation and sustainability value.

Post-it
Post-it are available in the worksheet for new concepts to stick new ideas, or ideas from the Sustainable Energy for All Idea Tables contained both in a dedicated file and in the Sustainable Energy for All section of the SDO toolkit tool.

How to use the tool
The Innovation Diagram for S.PSS&DRE tool requires the use of a slideshow software (e.g. Microsoft PowerPoint, or the equivalent in Open Office) or can be used in the printed version. According to the aim of the design activity, the corresponding worksheet/s need to be used.

How to analyse existing or competitor's energy offers:
First, write proposer/s name/s of who is working on it and the unit of satisfaction (e.g. access to energy, in the rural area, for home use). Second, position the existing and the competitor's offers (in the two worksheets) in the polarity diagram according to its customer and offer boundaries. As general rule is not compulsory that the offer correspond to a single position (e.g. B2C–B2B), if the case, is possible to locate the offer in the middle. Third, fill the profile following the instructions provided to copy/paste the labels. Considering that an existing or competitors' offer

is not automatically an S.PSS or is not necessarily offering products and/or services, some spaces in the profile could remain empty. Finally, write a short description of the offer emphasising innovation and sustainability problems.

How to design S.PSS applied to DRE concept
First, write the (draft) title of the concept that is going to be designed, then write the proposer/s name/s of who is working on it and the unit of satisfaction to be met (e.g. access to energy, in the rural area, for home use). Second, copy and paste the most promising ideas from the Sustainable Energy for All Idea Tables tool and position them in the polarity diagram. Creative discussions among the proposers will address the way to position the ideas according to customer and offer boundaries (the two polarity axes). As general rule is not compulsory that one idea corresponds to a single position (e.g. B2C–B2B), if the case, is possible to locate the idea in the middle and to decide after. Third, read all selected ideas and cluster them to create one/more concepts, some ideas if not interesting anymore can be excluded. Then, select the most promising S.PSS applied to DRE concept emerged and fill the profile following the instructions provided to copy/paste the labels. Finally, check coherence of the whole information and write the short description of the concept emphasising innovation and sustainability values. Follow up with discussion on the emerged S.PSS applied to DRE concept.

Integrating the tool into the design process
The Innovation Diagram for S.PSS and DRE can be used in the *Strategic Analysis* and *System concept design* stages of the design process.

Strategic Analysis
In the *Strategic Analysis,* it can be used to analyse and reorient existing energy offers, to analyse competitors' energy offers and even to make a comparison and start to identify potential opportunities.

System concept design
In the *System concept design,* it is used to combine the generated ideas and characterise the new S.PSS applied to DRE concept.

Results
The result in the case of existing or competitor's energy offers is their characterisation, where the lack of S.PSS applied to DRE offers emerge.

Tool availability and required resources
The tool is available for a free download at www.lenses.polimi.it. The tool has been designed to be used in workshops sessions, therefore, if the digital version is used a projector is preferable. In the case, the paper version is preferred suggestion is to print the worksheet as A3 or A2.

The time required to analyse existing or competitor's energy offers is approximately 20 min; in the case of the design of an S.PSS applied to DRE concept is approximately 30 min.

7.2.9 *Concept Description Form for S.PSS and DRE*

Aims

Concept Description Form for S.PSS and DRE [13] is a tool to visualise and finalise the description and characterization of a new S.PSS applied to DRE concept. The Concept Description Form presents a worksheet where to visualise key information, facilitating a deep understanding of the concept while presenting it among (existing—potential) stakeholders (Fig. 7.34).

What it consists of

The tool is composed of one worksheet with the following fields: proposer, title, unit of satisfaction, short description, profile.

Proposer the name/s of who is using the tool.

Title the name of the concept that is visualised with the tool.

Unit of satisfaction is the need satisfied/to be satisfied (e.g. access to energy, in the rural area, for home use).

Short description the short description is no more than 200 characters, to present the concept highlighting the main innovation and sustainability value.

Profile the profile presents a table with spaces to be filled with text on key information as: customer, provider, type of S.PSS, offered products (and related ownership), offered services (and related provider), what is paid, DRE system configuration, DRE source.

Fig. 7.34 Concept description form for S.PSS and DRE. *Source* designed by the Authors

How to use the tool
The Concept Description Form for S.PSS and DRE requires the use of a slideshow software (e.g. Microsoft PowerPoint, or the equivalent in Open Office) or can be used in the printed version. First, is needed to write proposer/s name/s of who is working on the concept, together with title and the unit of satisfaction met. Second, write the short description of the designed S.PSS applied to DRE concept emphasising innovation and sustainability values. Third, fill the profile table with text for each key information. Follow up with discussion on the emerged S.PSS applied to DRE concept and refine as needed. Generally, if the Innovation Diagram for S.PSS and DRE have been used, most information can be taken there and updated according to the newest version of the concept.

Integrating the tool into the design process
The Concept Description Form for S.PSS and DRE can be used in the *System concept design* stage of the design process. It is used also to present (internally and externally) the S.PSS applied to DRE concept.

Results
The result is the summary of an S.PSS applied to DRE concept, facilitating the concept definition while presenting it among (existing—potential) stakeholders.

Tool availability and required resources
The tool is available for a free download at www.lenses.polimi.it. The tool has been designed to be used in workshops sessions, therefore, if the digital version is used a projector is preferable; in the case, the paper version is preferred suggestion is to print the worksheet as A2 or A1. The time required to summarise an S.PSS applied to DRE concept is approximately 20 min.

7.2.10 Stakeholder Motivation and Sustainability Table

Aims
The Stakeholders Motivation Matrix [10], a tool to visualise motivations of the stakeholders, has been updated [13] as a collaboration between DIS Research Group of Politecnico di Milano (Italy), Makerere University (Uganda) and TU Delft University (The Netherlands) becoming Stakeholders Sustainability and Motivation Table. It is presented as visualisation tool aimed to identify/show: motivations and contributions of each stakeholder; sustainable (economic, environmental, socioethical) benefits from each stakeholder; this facilitating involvement process and strategic conversations addressing various (existing and potential) stakeholders (Fig. 7.35).

What it consists of
The tool is made of four worksheets: the table, two worksheets with guidelines to define environmental and socioethical benefits, a worksheet with icons.

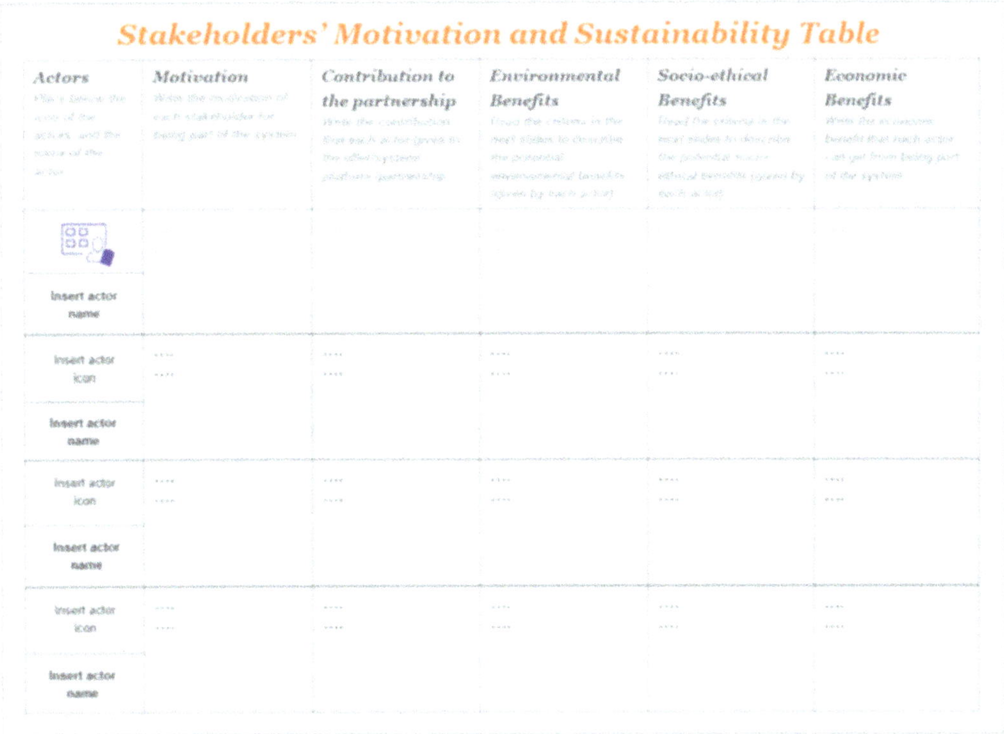

Fig. 7.35 Stakeholders motivation and sustainability table. *Source* designed by the Authors

Table The table worksheet is made of a table with six columns: stakeholders, motivation, contribution to the partnership, environmental benefits, socioethical benefits, economic benefits and many lines according to number of stakeholders.

Worksheets with checklists these worksheets present environmental and socioethical checklists to address the definition of sustainable benefits by each stakeholder.

Worksheet with icons this worksheet presents icons representing several possible stakeholders, divided as providers and customers, that can be used in the first column of the table to describe each stakeholder.

How to use the tool

For each stakeholder is asked to fill all columns: stakeholders: stakeholder icon and stakeholder name; motivation: motivations for the specific stakeholder to be in the partnership of stakeholders/contribution to the partnership—contribution given by the stakeholder to the partnership; environmental, socioethical, economic benefits—benefits brought from the specific stakeholders in relation to sustainability. Follow up with a preliminary discussion addressing (existing and potential) stakeholders. To fill the environmental and socioethical benefits two dedicated worksheets are available.

Integrating the tool into the design process

The Stakeholders Motivation and Sustainability Table can be used in the *System concept design* and *Design System Details* stages of the design process.

In both cases, it can be used to verify and facilitate the involvement process and to orient strategic conversations addressing (existing and potential) stakeholders.

Results
The result is an informative table of motivations, contributions and potential benefits as way to orient strategic conversations addressing (existing and potential) stakeholders.

Tool availability and required resources
The tool is available for a free download at www.lenses.polimi.it. The Stakeholders Sustainability and Motivation Table requires the use of a slideshow software (e.g. Microsoft PowerPoint, or the equivalent in Open Office) or can be used in the printed version, in this case, printed materials and a pen are sufficient. The tool has been designed to be used in workshops sessions, therefore, if the digital version is used a projector is preferable; in the case, the paper version is preferred suggestion is to print the worksheet as A3 or A2. The time required to fill the information is approximately 10 min for each stakeholder.

References

1. Bacchetti E, Vezzoli C, Vanitkoopalangkul K (2016) Sustainable product-service system applied to distributed renewable energies: a scenario tool. In: Vezzoli C, Delfino E (eds) Sustainable energy for all by design conference proceedings, pp 157–166
2. Emili S, Ceschin F, Harrison D (2016a) Product-service systems applied to distributed renewable energy: a classification system and 15 archetypal models. Energy Sustain Develop 32:71–98
3. Emili S, Ceschin F, Harrison D (2016b) Supporting SMEs in designing product-service systems applied to distributed renewable energy: design framework and cards. In: Proceedings of the LeNSes international conference 2016, pp 28–30, Cape Town, Sept 2016
4. Emili S, Ceschin F, Harrison D (2016c) Design-supporting tools for visualising product-service systems applied to distributed renewable energy: the energy system map. In: Proceedings of the LeNSes international conference 2016, pp 28–30, Cape Town, Sept 2016
5. Emili S, Ceschin F, Harrison D (2016d) Supporting SMEs in designing sustainable business models for energy access for the BoP: a strategic design tool. In: Conference proceedings of design research society (DRS2016), pp 27–30, Brighton, UK, June 2016
6. Emili S, Ceschin F, Harrison D (2015e) Product-service systems applied to distributed renewable energy systems: a classification system and a strategic design tool. In: Conference proceedings KIDEC—Indigenous Design—Kampala 2015
7. Emili S (2017) Designing Product-Service Systems applied to Distributed Renewable Energy in low-income and developing contexts: A strategic design toolkit. PhD Thesis, Brunel University London
8. Kiravu C, Emili S., Magole L, Jeffrey A, Mbekomize C, Matlotse E, Rakgati E, Oladiran T, Tsamaatse K, Ceschin F (2015) Mmokolodi solar PV project—demonstrating sustainable renewable energy system design and potential rural electrification in Botswana. In: International conference on clean energy for sustainable growth in developing countries 2015, pp 16–18, Palapye, Botswana, Sept 2015
9. Jégou F, Manzini E, Meroni A (2002) Design plan, a tool for organising the design activities oriented to generate sustainable solutions, Working paper, SusProNet conference, Amsterdam

10. Jégou F, Manzini E, Meroni A (2004) Design plan. A design toolbox to facilitate solution oriented partnerships. In: Manzini E, Collina L, Evans S (eds) Solution oriented partnership. Cranfield University, Cranfield
11. Manzini E, Jégou F, Meroni A (2009) Design-oriented scenarios. In: Crul M, Diehl JC (eds) Design for sustainability (D4S): a step-by-step approach. Modules, United Nations Environment Program (UNEP), pp 15–32
12. Vezzoli C, Bacchetti E (2017) The sustainable energy for all design scenario. In: Chapman Jonathan (ed) The Routledge handbook of sustainable product design. Routledge, New York, pp 443–464
13. Vezzoli C, Bacchetti E, Ceschin F, Moalosi R, Nakazibwe V, Osanjo L, M'Rhitaa M, Costa F (2016) System design for sustainable energy for all. A new knowledge base and know-how developed within the LeNSes European and African project. In: Vezzoli C, Delfino E (eds) Sustainable energy for all by design conference proceedings, pp 95–110

Chapter 8
Practical Examples of Application of SD4SEA Approach/Tools

8.1 Introduction

This chapter illustrates two practical applications of the SD4SEA design approach and tools, describing how they have been used by companies, practitioners and academics in different countries as part of the LeNSes project.

The tools have been applied in practice with different types of users. On the one hand, companies and practitioners (NGO, consultants, and designers) used the tools for a range of purposes ranging from understanding the market in a given geographic area to exploring new sustainable business opportunities to design concepts of S.PSS applied to DRE. On the other hand, academics and teachers used the SD4SEA approach and tools to teach the various aspects of designing and developing S.PSS applied to DRE.

The following sections describe two cases of application of the SD4SEA design approach and tools.

- ***Case 1:*** *Solar energy company (Botswana)*
  Tools used: S.PSS & DRE Innovation Map, the S.PSS & DRE Design Framework and Cards, Energy System Map;
  Objectives: to explore new business models and other technology options in order to reach a wider range of customers in Botswana.
- ***Case 2:*** *SMEs for energy (Uganda)*
  Tools used: Innovation Diagram for S.PSS&DRE, Sustainability Design Orienting Scenario for S.PSS&DRE, Sustainable Energy for All Idea Tables (and cards), Energy System Map, Stakeholders Sustainability and Motivation Table;
  Objectives: to innovate and increase sustainability of the current business of the SMEs, adopting the Sustainable Product-Service System applied to Distributed Renewable Energy model.

C. Vezzoli et al., *Designing Sustainable Energy for All*, Green Energy and Technology, https://doi.org/10.1007/978-3-319-70223-0_8

8.2 Solar Energy Company, Botswana

Context and Objectives

An example of how a company used some of the S.PSS and DRE design tools is related to a workshop organised to support SMEs in developing sustainable Product-Service Systems for energy access in African contexts.

A company from Botswana was involved in a three-day workshop to redesign their business model. The company sells mini kits and solar products with consultancy and training services. They aimed at expanding their portfolio of offerings to other customers and possibly including new products in their range. After a short introduction on S.PSS and DRE models, their benefits and the proposed design approach, participants used some of the SD4SEA tools to refine and re-orient their business model.

Description of Activities

1. **Exploring the applications of S.PSS and DRE in low-income and developing contexts.**

Participants were first introduced to S.PSS applied to DRE, their benefits and the design tools. Then, they used the Innovation Map and the Archetypal Models cards to map five examples of case studies on the map, positioning them according to the S.PSS type, the energy system used and the target user. This activity aimed at getting familiar with the Innovation Map and at understanding different types of S. PSS and DRE offers.

2. **Strategic analysis with the Innovation Map**

The company started this task by positioning their current offerings on the Innovation Map according to the type of energy system, the target customer and the S.PSS type. The company positioned themselves on the quadrant related to 'pay-to-purchase mini kits with advice and consultancy services'. Looking at other options provided by the Map, they immediately thought about moving towards leasing models and pay-per-unit of satisfaction of both mini kits and bigger individual solar systems. These discussions were triggered by the fact that most competitors in the market are operating in the 'pay-to-purchase' area. In fact, during the discussion on competitors, participants positioned all of them in the bottom part of the Map and on the Non-PSS offers area. Because of these reasons they decided to explore types of offers that were not provided in the context of Botswana (see Fig. 7.35).

3. **Concept generation with the Innovation Map**

In a second phase, participants used the tool to brainstorm about new concepts. They used the Concept Cards to define three new business models and then position them on the corresponding area of the map. As illustrated in Fig. 2.5, concepts were composed by a combination of different offers. Concept 1 combines a product-oriented offer (pay-to-purchase with additional services) with a use-oriented one (leasing model) involving solar water pumps offered through an entrepreneur-managed model. Concept 2 involves the provision of energy services through solar mini kits on a pay-per-unit of satisfaction. Concept 3 combines a use- and a result-oriented S.PSS and involves leasing charging stations (solar kiosks) to mobile money producers, employing local entrepreneurs to provide charging services to end-users.

While Concept 1 employs the solar mini kits technology, which corresponds to the current type of products offered by the company, the other concepts involve larger systems and charging stations. In fact, after having mapped a competitor providing solar kiosks in the non-PSS area, the company brainstormed about possible partnerships to set up with this company, with the aim of reaching a wider number of customers.

Another interesting aspect emerging from this first idea generation was the decision to target different types of users. The company, in fact, identified areas for opportunities in the farming sector and in off-grid communities, brainstorming about different technology options to satisfy their energy needs (solar water pumps and charging stations) (Fig. 8.1).

4. **Concept detailing with the Design Framework and Cards**

The second day of the workshop focused on detailing the concepts generated by using the Design Framework and Cards (see Sect. 7.2.7). They were given the

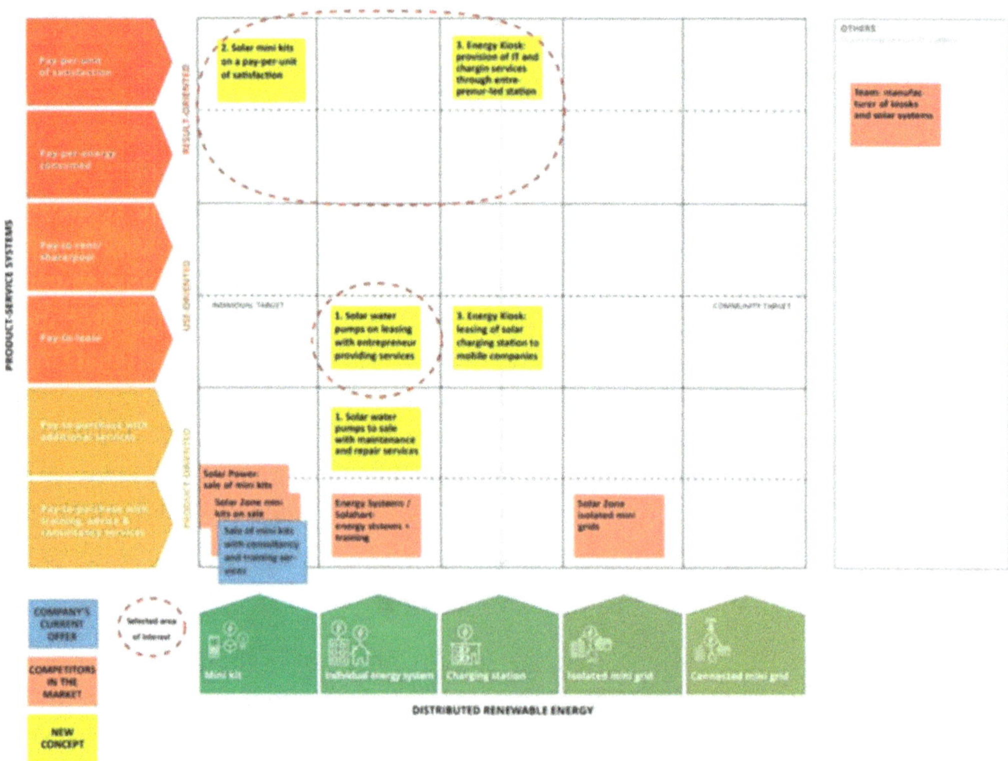

Fig. 8.1 Innovation map completed by the SME in Botswana. *Source* designed by the Authors

Framework with Cards and a Design Canvas to be filled out with post-its. By browsing the Cards and getting inspiration from case studies and guidelines, participants completed their own Design Canvas (Fig. 7.32).

After having completed the first idea generation with the Innovation Map, the company had to decide which concepts to select for the detailing phase. The company initially decided to focus on result-oriented S.PSS (pay-per-unit of satisfaction) for their mini kit concept, positioned in the corresponding area of the Map (Fig. 2.5). However, after having discussed implications for implementing this model and necessary resources needed (such as capital financing), they decided to return to their initial business offer (offering mini kits on a pay-to-purchase with additional services) and kept the result-oriented model as a concept idea to be implemented in future. This suggests that the Innovation Map helped the company in identifying and detailing new strategic opportunities to be pursued in future, even though these cannot be implemented straight away.

The brainstorming session was then focused on developing all three concepts selling mini kits with consultancy services; providing solar water pumps on leasing and on sale to farmers; providing charging stations on leasing to entrepreneurs who would then provide charging services to end-users (pay-per-unit of satisfaction). To avoid confusion, ideas were written down on different types of post-it (Fig. 2.6).

This activity helped the company in detailing the network of stakeholder involved (partnership with local manufacturer and local entrepreneurs) and in understanding the different services they would need to integrate in their offers. In particular, they included installation, maintenance, as well as training on product management targeted to local entrepreneurs. The company also discussed about providing end-of-life services and collection of extinguished batteries, a service that currently no other actor offers in Botswana (Fig. 8.2).

5. **Visualisation and communication with the Energy System Map**

The last phase of the workshop focused on using the Energy System Map (see Sect. 7.2.7) to detail some aspects on the new solutions and to visualise the entire model. Participants were provided with a printed example of the tool, a set of icons and a template to use for designing their own system map. By cutting the icons and pasting them on the template, participants identified the main elements of their business model. In the second stage, they drew flows of information, services, goods and money between stakeholders (Fig. 2.7). The company affirmed that this process helped them in clarifying some aspects of their concepts, especially in terms of payment flows. In fact, using the tool at the end of the idea generation session helped them in identifying issues in their concepts and overall achieving a higher level of detail (Fig. 8.3).

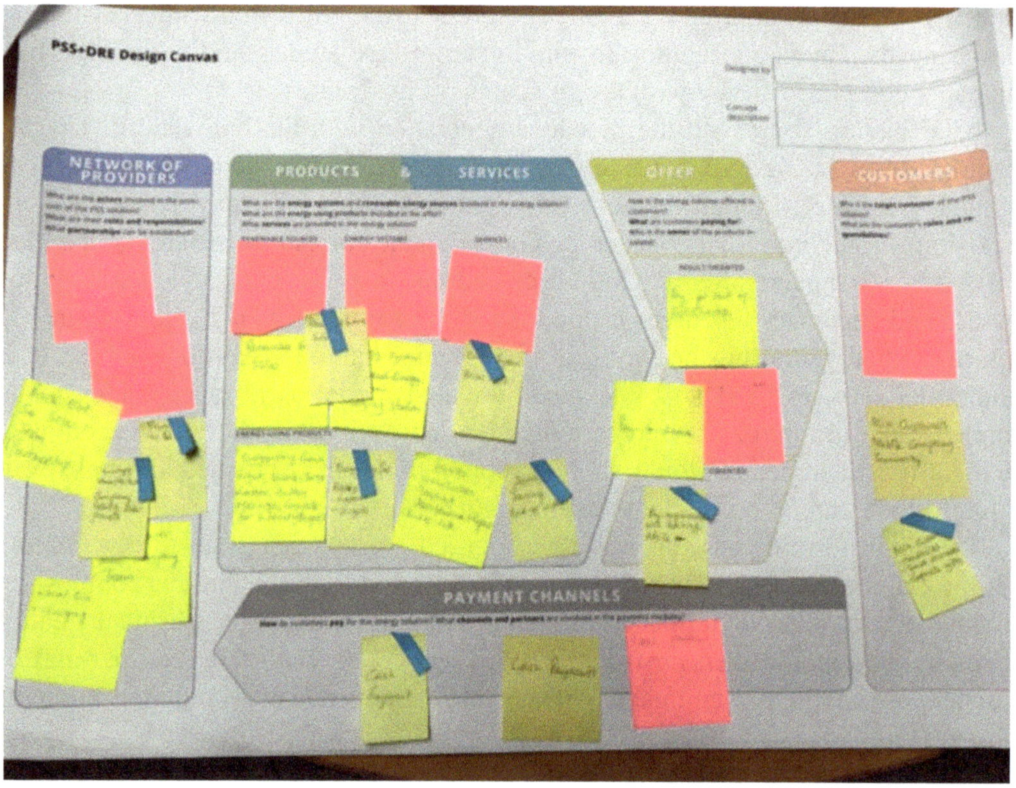

Fig. 8.2 The design canvas produced by the company. *Source* designed by the Authors

Outcomes

The company currently sells solar mini kits with consultancy services. After having applied some of the SD4SEA tools, the company explored the opportunity of shifting their current offerings on different types of S.PSS offers, exploring different technology options and target customers. In terms of offering, they generated concepts in the use and result-oriented areas, moving away from the product-oriented area where they currently operate. Moreover, the company combined two models, leasing option and pay-per-unit of satisfaction, in their solar charging station concept (energy kiosk).

This example illustrates how companies can design solutions moving away from their current product-oriented models towards ownerless-based offers. According to feedback received by the company, the tools helped them in identifying opportunities for their chosen market and a promising niche to explore (*'it was helpful to see where this niche markets are amongst competitors. It gives a good visualisation of where the current market is heading... you are able to take advantage of opportunities not being explored'*). In fact, the company was able to see that all competitors in Botswana are located in the product-oriented area, and thus that interesting opportunities to create a competitive advantage lie in providing use- and result-oriented S.PSSs.

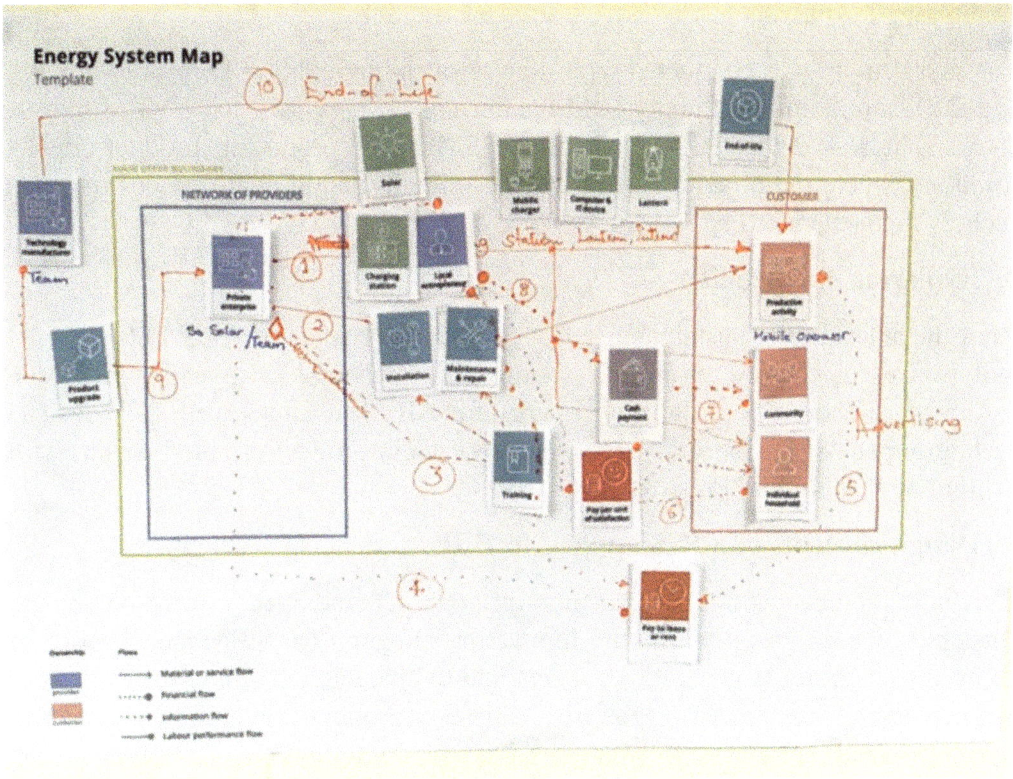

Fig. 8.3 The energy system map produced by the company. *Source* designed by the Authors

8.3 SMEs for Energy, Uganda

A further prototyping of the SD4SEA tools was conducted by the Makerere University (Uganda—2016) as a collaboration between the Centre for Research in Energy and Energy Conservation (CREEC) of the University and Politecnico di Milano. The course involved nine Small and Medium Enterprises (SMEs) for energy from Uganda.

Objectives
Participants were asked to innovate and increase sustainability of their existing businesses, by designing Sustainable Product-Service System applied to Distributed Renewable Energy concepts. Attention was addressed to designing new concepts, and to properly communicate them to external audiences using dedicated tools.

Description of Activities
The SMEs representatives were asked to work in groups of 3–4 practitioners, dealing with different Distributed Renewable Energy (DRE) such as biogas, sun, hydropower and cook-stove technologies. The course was based on theoretical lectures, case studies and a design consultancy.

1. **Strategic Analysis of the SMEs state of the art**

The first activity was conducted with the use of the Innovation Diagram for S.PSS and DRE tool, aiming to understand the current business of each SME. From the analysis, it was evident that most of the SMEs are proposing product-oriented solutions, where the product is sold with (eventually) additional services included, such as maintenance (Fig. 8.4).

2. **Exploring opportunities**

After the analysis, the Sustainability Design Orienting Scenario for S.PSS and DRE tool was used to show promising visions (four videos), to give inspirations to participants. Then, the Sustainable Energy for All Idea Tables were used. In fact, each group designed several ideas to move their product-oriented business, to explore new solutions (Figs. 8.5 and 8.6).

3. **Design concepts of S.PSS applied to DRE**

The most promising system ideas among those generated, they were copied and clustered by each group within the Innovation Diagram for S.PSS and DRE. This allowed each group to generate a concept and to characterise it in terms of network of providers, customer/s, type of S.PSS (Product-oriented, Use-oriented, Result-oriented), products and services offered, configuration of the system and type/s of renewable resources. To clarify the interactions of (potential) actors of the system, the Energy System Map tool was used by all groups, and as well the

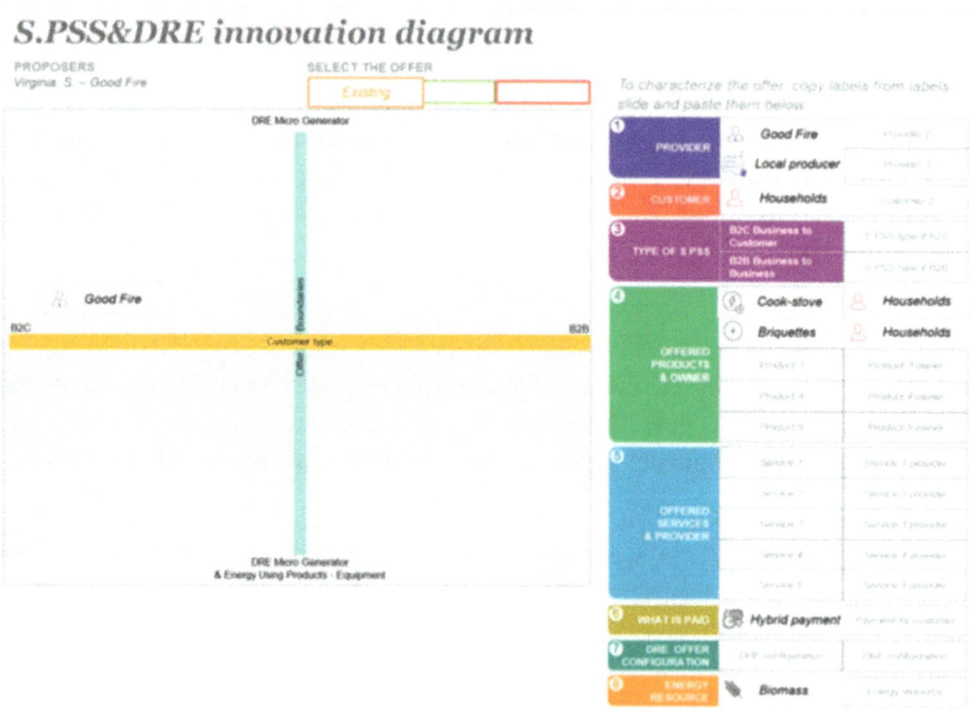

Fig. 8.4 Current business of a SME involved in the course. *Source* designed by the Authors

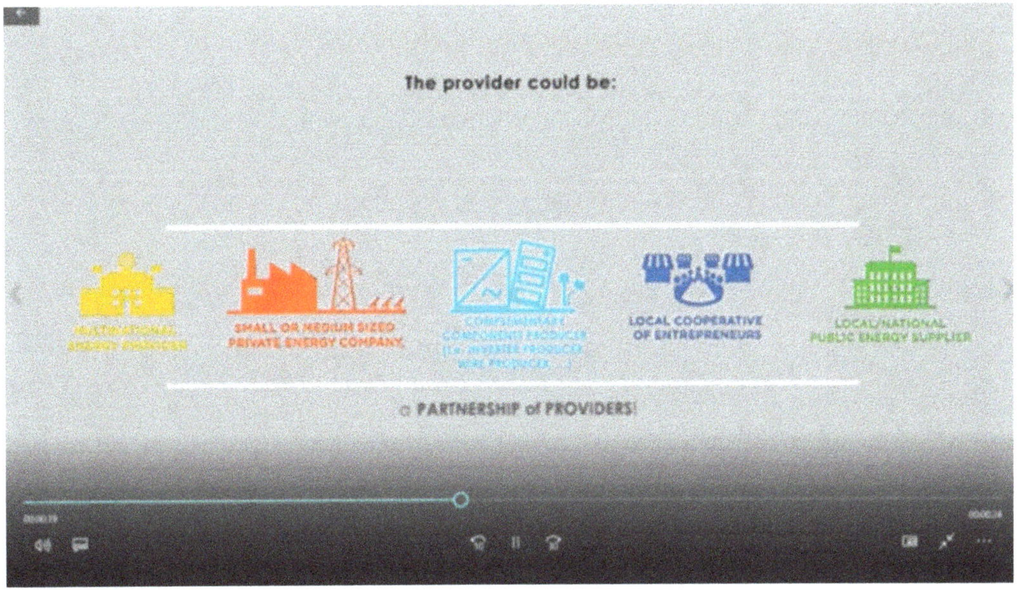

Fig. 8.5 Screenshot from sustainability design orienting scenario for S.PSS and DRE. *Source* designed by the Authors

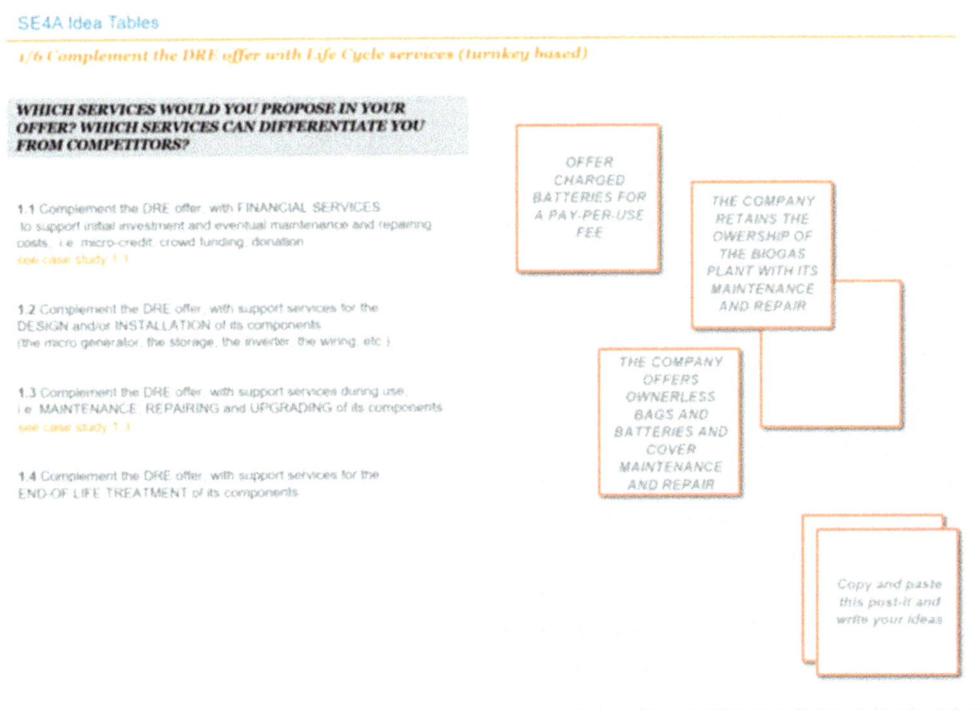

Fig. 8.6 Ideas generated using the SE4All idea generation tables and cards. *Source* designed by the Authors

Fig. 8.7 Stakeholders' motivation and sustainability table generated by participants. *Source* designed by the Authors

Stakeholders' Motivation and Sustainability Table tool, which brought more details on motivations/contributions/benefits from and for each of the (potential) stakeholders (Fig. 8.7).

Outcomes

Three concepts of S.PSS applied to DRE were developed, thus opening innovative opportunities for their current businesses. One of the concepts was *'A business to customer (B2C) solution, based on a community bio-digester, owned by the Renewable Energies Ltd (REL), who is responsible for its installation, training, repair and maintenance. REL offers to its customers biogas stored in bags to facilitate cooking activities and charged batteries for lanterns. Customers pay-per-use to use the energy services (biogas refill/battery charging). REL owns biogas bags and the batteries, customers own the stoves and the lights. To gain extra-money and Customers can provide bio-waste to support the function of the bio-digester, that will be paid from REL'.*

8.4 Summary and Considerations

The SD4SEA tools, approach and support have been used (and tested) not only in the above-described situations. In all 10 organisations (SMEs, NGOs, Research Centres) and 10 students and 10 teachers have been involved in courses and lifelong learning modules. In fact, the tools have been applied by companies, practitioners and students in four African countries and in Europe. The experiences conducted validated the tools and their adaptability to different purposes of application.

All feedbacks have been very positive, thus encouraging the diffusion and their use in both low, middle and high-income contexts.